D0906553

QUANTITATIVE ANALYSIS
BY GAS CHROMATOGRAPHY

CHROMATOGRAPHIC SCIENCE

A Series of Monographs

VOLUME 1 : Dynamics of Chromatography (in three parts), *J. Calvin Giddings*

VOLUME 2 : Gas Chromatographic Analysis of Drugs and Pesticides, *Benjamin J. Gudzinowicz*

VOLUME 3 : Principles of Adsorption Chromatography; The Separation of Nonionic Organic Compounds, *Lloyd R. Snyder* (out of print)

VOLUME 4 : Multicomponent Chromatography; Theory of Interference, *Friedrich Helfferich and Gerhard Klein* (out of print)

VOLUME 5 : Quantitative Analysis by Gas Chromatography, *Josef Novák*

OTHER VOLUMES IN PREPARATION

QUANTITATIVE ANALYSIS
BY GAS CHROMATOGRAPHY

JOSEF NOVÁK

INSTITUTE OF ANALYTICAL CHEMISTRY
CZECHOSLOVAK ACADEMY OF SCIENCES
BRNO CZECHOSLOVAKIA

MARCEL DEKKER INC.   New York and Basel

MARCEL DEKKER, INC.
270 Madison Avenue, New York, New York 10016

LIBRARY OF CONGRESS CATALOG CARD NUMBER: 75-10349

ISBN: 0-8247-6311-4

Current Printing (last digit):
10 9 8 7 6 5 4 3 2

PRINTED IN THE UNITED STATES OF AMERICA

CONTENTS

Preface   vii

Chapter 1   INTRODUCTION   1

Chapter 2   CONCENTRATION OF THE SOLUTE COMPONENT IN
             THE ELUTED CHROMATOGRAPHIC ZONE   7

   I.   Mean Concentration   8
  II.   Zone-Maximum Concentration   10

Chapter 3   DETECTION   13

   I.   Definitions of Fundamental Terms   13
  II.   Classification of Detectors   14
 III.   Performance Characteristics of Model Types
        of Detector   19

Chapter 4   RELATIONS BETWEEN PEAK AREA AND THE
             AMOUNT OF SOLUTE COMPONENT IN THE
             CHROMATOGRAPHIC BAND   25

    I.   Analysis of the Signal and Response-
         Determining Parameters   25
   II.   Linearity of Response   29
  III.   Relations Between the Instantaneous Amount
         of the Substance Chromatographed in the
         Sensing Element and the Detector Response   31
   IV.   Relations Between the Total Amount of the
         Substance Chromatographed Passed Through
         the Sensing Element and the Time Integral
         of the Detector Response   34
    V.   The Role of an Auxiliary Gas   36
   VI.   Molar and Relative Molar Response;
         Correction Factors   40
  VII.   Analytical Significance of Uncorrected
         Quantitative Parameters of the Chromatogram   44

Chapter 5   PREDICTION OF THE RELATIVE MOLAR
             RESPONSE   47

    I.   Katharometer   47
   II.   Martin's Gas-Density Balance   51
  III.   Scott's Microflame Detector   53
   IV.   Flame-Ionization Detector   55
    V.   Cross-Section Ionization Detector   58
   VI.   Electron-Capture Detector   60
  VII.   Argon Ionization Detector   63

Chapter 6   CONVENTIONAL TECHNIQUES   67

    I.   Problem Analysis   68
   II.   Presentation of Concentration   71
  III.   Survey of Working Techniques   74
   IV.   Sample Dilution   96
    V.   Experimental Determination of
         Correction Factors   104

Chapter 7   SPECIAL TECHNIQUES   107

    I.   Trapping of Components of Gaseous Samples   108
   II.   Extraction of Liquid Samples by
         Liquid Extrahents   117
  III.   Head-Space Gas Analysis   138
   IV.   Liquid Extraction and Head-Space Gas
         Analysis of Nonfluidic Materials   153

Contents   v

Chapter 8   QUANTITATION OF THE CHROMATOGRAM   157

    I.   Manual Techniques of Peak-Area Determination   157
   II.   Methods of Peak-Area Calculation   159
  III.   Interpretation of Chromatograms by Peak Heights   161
   IV.   Automatic Processing of Chromatograms   164
    V.   Resolution of Overlapped Peaks   172

Chapter 9   RELIABILITY OF RESULTS   177

    I.   Fundamental Statistical Definitions   178
   II.   Sources of Errors   180
  III.   Precision of Results Obtained by
         Conventional Techniques   183
   IV.   Precision of the Quantities Representing
         Independent Variables   195

References   199

Index   209

PREFACE

Traditional quantitative gas chromatography has hitherto
been conceived mainly from a practical point of view. (The
term "traditional" is used to indicate that those problems are
meant which are associated methodically with the traditional
instrumentation of gas chromatography, rather than the problems
of computer-aided processing of chromatograms.) However, this
conception skews the picture on the significance of quantitative gas chromatography and affords no more than a cursory
understanding of this important discipline. Actually, quantitative gas chromatography represents the only analytical variant of gas chromatography that per se provides unambiguous
results; in qualitative gas chromatography, unequivocal identification of components can be obtained only through combination with an independent analytical method.

The scope of quantitative gas chromatography is substantially broader than that of gas chromatography alone; in addition to the problems of the separation process, it is necessary to manage also the problems of the numerous gas chromatographic detectors and the arrangements for processing their responses. Thus, quantitative gas chromatography has associated with it a number of new conceptual qualities which cannot be done justice in a mere chapter of a book on the general subject of gas chromatography. At the present state of development of gas chromatography, quantitative gas chromatography has to be understood as a self-contained modern analytical discipline the theory of which is built up on the mastered principles of general gas chromatography. It is believed that this subject deserves a monograph of this size.

The book is designed primarily for those who have some acquaintance with gas chromatography, but it is hoped that it will prove useful also to the beginner in the field. Chapters 1-5 (about a third of the book) deal with the basic theory of quantitative gas chromatography. The goal of this part of the book is to demonstrate the properties of gas chromatographic detectors, present accurate relations between the integral detector response and the amount of solute component in the chromatographic zone, and show the possibilities of predicting the detector-response factors. Chapters 6 and 7 (also about a third of the book) concern the methodological aspects of the problems. Chapter 6 presents a consistent survey of conventional working techniques. Chapter 7 is devoted to special techniques, involving matters such as isolation of trace components from ballast material, accumulation of trace components, and analysis of multiphase systems. Special attention is given to quantitative head-space gas analysis; based on the standard addition technique combined with the mass-balance of solute in a two-phase system, methods have been formulated which enable

the determination of the content of a component in the whole
system to be carried out by virtue of the analysis of only one
of the phases. However, only general concepts of these methods
have been defined, and there is much left to be done in the
elaboration of procedures suitable for particular analytical
problems. Chapter 8 describes the problems and procedures of
processing chromatograms, and Chapter 9 deals with the analytical errors encountered in quantitative gas chromatography. The
problems of automatic processing of chromatograms are discussed
briefly in Chapter 8.

   The material in this book has grown and matured since about
1960. It is a pleasant duty to thank Dr. Jaroslav Janák, who
has created a unique atmosphere for this work and for my pro-
fessional growth and was a good counsel at all times. The
method of equilibrium concentration of trace components has
arisen from enjoyable collaboration with Dr. Vladimír Vašák.
At a later stage of the development of this material, Dr. Peter
Boček was a significant participant in a trial on the reliability
of conventional techniques of quantitative gas chromatography.
The section on the automatic processing of chromatograms is
largely due to Dr. Stanislav Wičar.

   Despite the care it has received, the book will certainly
have shortcomings. However, nothing is perfect in the world
of reality, and anything that should be made absolutely ideal
could never be finished. I shall appreciate any comments from
readers.

June, 1975   Josef Novák
Brno, Czechoslovakia

QUANTITATIVE ANALYSIS
BY GAS CHROMATOGRAPHY

Chapter 1

INTRODUCTION

Quantitative analysis by gas chromatography is a very broad problem that can be taken from several viewpoints and approached in various ways. This book is devoted primarily to the theoretical aspects of this problem; the main goal is to show that quantitative analysis by gas chromatography can be looked upon as a self-contained analytical discipline having its own consistent theory.

With the aim of making it possible to define precisely the concept of this discipline, a boundary was deliberately laid between the area associated directly with the chromatographic system proper and the area that can be considered to be more or less independent of this system. Under the term chromatographic system the arrangement is meant which corresponds to the concept of a standard analytical gas chromatograph,

1

comprising the subsystems such as gas-flow control, column, column-temperature control, and detector with the necessary accessories. Hence, the first area covers those quantitative-analytical aspects of gas chromatography which are directly related to the properties of the subsystems. The second area comprises the problems associated with the methods of instrumental integration of the detector response and computer-based evaluation of the results of analysis. Although these methods are of extremely great significance in quantitative analysis by gas chromatography, and the respective instrumentation becomes gradually a standard outfit of the modern GC laboratory, their theoretical background lies in the domain of the techniques of computerization rather than in gas chromatography.

While the material pertaining to the first area has been continuously developed and compiled since the very advent of gas chromatography, the second area began to evolve later. Thus, the problems of the first area are ripening gradually, whereas those of the second follow the rapidly advancing development of computerization techniques. In this book, the primary focus is the first area; the problems of the second area are discussed only briefly, in order to make the contour of quantitative gas chromatography complete.

In the foregoing context, quantitative analysis by gas chromatography can be defined as a method comprising the separation on a GC column of an n-component mixture to produce n binary solute/carrier-gas mixtures, and the on-line determination of the solutes in these mixtures by a special analyzer, the GC detector. The mixtures are represented by the individual chromatographic zones. Owing to the nature of the chromatographic process, the instantaneous solute concentration in the column effluent displays a certain time dependence during the elution of the chromatographic zone, which must be closely followed by the detector. The magnitude of the record of a

chromatographic zone, regardless of whether it has been provided in an analog or digital form, is proportional to the concentration or mass of the solute substance in the zone.

It is only in some cases involving the use of integral
detectors that the overall detector responses (differences
between the responses at the end and the beginning of elution)
are related directly to the solute amounts in the individual
chromatographic zones and that the necessary proportionality
constants can be determined a priori by virtue of the chemical
reactions that are taking place in the process of detection.
A typical example thereof is the pioneer work by James and
Martin [1]. Still more straightforward cases are those involving the procedures described by Janák [2] and by Bevan and
Thornburn [3]; in these cases, the overall response is literally identical to the volume or mass of the solute substance
in the zone.

Unfortunately, in the majority of cases usual in gas chromatography, the situation is much more complicated. The most
important detectors are those of differential type, which renders it necessary to integrate the response in order to obtain
univocal quantitative data. Further, like the qualitative
data, which are to a certain extent stigmatized by quantitative
factors, the quantitative characteristics are dependent on the
quality of the material chromatographed. Examples are the dependence of retention characteristics on the sample size
(dependence of the partition coefficient on the concentration
of solute in the sorbent) on the one hand [4] and, on the other
hand, the well-known variability in the sensitivity of a particular detector to various types of material. Such effects
are the usual sources of undesirable modifications of the
respective data and have to be eliminated. Only in special
cases can these secondary effects be utilized to advantage for
analytical purposes. Examples of the latter are to be found

in the use of selective detectors, e.g., electron capture [5],
alkali flame ionization [6], and coulometric detectors [7],
which make it possible to achieve quantitative determination
and group identification simultaneously. From the viewpoint
of quantitative analysis, of course, any detection specificity
calls for the introduction of corrections. This creates the
problem of defining the correction factors and the problem of
concentration units for expressing the results. A reliable
orientation in these problems can be gained only on the basis
of theoretical analysis.

Because of its inherent relationship with the problems of
detection, quantitative gas chromatography reaches into several
fields based on theoretical foundations entirely different from
those of the separation process. This fact and the outpacing
of the theory by practice, which has been prompted by the un-
usual attractiveness of gas chromatography, have resulted in
some vagueness, especially as concerns the corrections of
quantitative data of the chromatogram and the concentration
units to be used in expressing the results. Usual causes of
this vagueness have been, as a rule, unjustifiable generalizations made from empirical observations. Thus, for example,
one can find in the literature [8] a statement which claims
that the problem of concentration units is a matter which had
been solved, with reference to the paper by Hausdorff [9], who
found that in the gas chromatography of acetone and carbon
tetrachloride with hydrogen carrier gas and katharometer detection the uncorrected peak areas are proportional to the
mole percentage of the components. A theoretical analysis [10]
shows, however, that the postulate, which is quite correct in
this specific case, cannot be generalized; both theory and
practice [11-18] have clearly shown that the proportionality
of uncorrected quantitative data to weight percentage of the
respective components has more general validity.

Quantitative analysis by gas chromatography imposes high requirements for instrumentation and necessitates meeting high demands on the sampling technique, sample adjusting and dosing, and the evaluation of the chromatogram. These aspects are quite general in nature, as discussed by Evans and Scott [19]. The relations between the properties of the detection and recording systems and the distortion of the chromatographic record were studied in detail by Sternberg [20], Hána [21], and McWilliam and Bolton [22].

In the existing monographs on gas chromatography, little space has been devoted to the problems of quantitative analysis. Their respective chapters give merely brief descriptions of the working techniques and the methods used for evaluating the chromatogram. The only book focused on the questions of quantitative gas chromatography is that by Kaiser [8]. A significant contribution to the field of quantitative gas chromatography has been Sternberg's study on detectors [23].

Gas chromatography can cover a considerable proportion of the problems of analytical chemistry. The scope of gas chromatography, with regard to the complexity of mixtures and character of material submitted to analysis, can be controlled within wide limits by the choice of the chromatographic system and operation conditions (capillary GC, temperature programming, high pressure, special modes of detection). The extent to which the versatility of gas chromatography can be utilized is determined by the existing degree of instrumentation refinement.

Chapter 2

CONCENTRATION OF THE SOLUTE COMPONENT IN THE ELUTED
CHROMATOGRAPHIC ZONE

After its introduction into the chromatographic column, the
solute component quickly distributes itself between the mobile
phase (m) and the sorbent (s). The distribution ratio is given
by the capacity ratio k defined by

$$k = \frac{N_{is}}{N_{im}} = K \frac{V_s}{V_m} \qquad\qquad (2\text{-}1)$$

where N_{is} and N_{im} are the number of moles of the solute (i)
present in the sorbent and in the mobile phase, K is the partition coefficient of the solute in the given system, and V_s and
V_m are the volumes of the sorbent and of the mobile phase in the
column; K is given by $K = (N_{is}/V_s)/(N_{im}/V_m)$. As long as the
chromatographic bed is homogeneous, the working conditions

constant, and the sorption isotherm linear, the capacity ratio does not vary along the migration path of the zone. Thus, for N_i moles of solute introduced into the column, we can write

$$N_i = N_{im} + N_{is} \tag{2-2}$$

which, on combining with Eq. (2-1), gives

$$N_{im} = \frac{N_i}{1 + k} . \tag{2-3}$$

I. MEAN CONCENTRATION

Let us consider a chromatographic zone migrating along the coordinate z (column axis), of instantaneous actual width Δz and containing N_{im} moles of solute i in the mobile (gas) phase. If the gas-phase cross-sectional area of the column is ϕ, the volume of the gas within the zone is $\Delta V_z = \phi \, \Delta z$, and the mean solute concentration (number of moles per unit volume) in the mobile phase of the zone is $c'_{iz} = N_{im}/\Delta V_z$. As the zone migrates down the column, it gradually broadens, i.e., the values of Δz and ΔV_z increase and, consequently, c'_{iz} decreases; this variability is indicated by subscript z. The quantity ΔV_z can further be expressed by $\Delta V_z = \phi \, \Delta t_z \, u_{iz}$, where u_{iz} is the forward velocity of the migration of the zone center and Δt_z is the time interval between the beginning and end of the passage of the zone through the column cross section at a distance z from the inlet. The ratio of the zone-center to mobile-phase velocities, u_{iz}/u_z, is the so-called retardation factor R; this can be expressed [24] by $u_{iz}/u_z = R = 1/(1 + k)$. The variability of u_{iz} and u_z is due to the compressibility of the carrier gas. As both u_{iz} and u_z vary in the same manner, the values of R and k are invariant. Hence, ΔV_z can be written as $\Delta V_z = \phi \, \Delta t_z \, u_z/(1 + k)$.

When considering the situation at the column outlet, then
z, Δz, and all the z-dependent quantities have fixed values
($z = \underline{L}$, $\Delta z = \Delta\underline{L}$, $\underline{L}$ being the column length), i.e., the subscript z can be omitted, and the mean solute concentration in
the column effluent at the column outlet, c_i', is

$$c_i' = \frac{N_{im}}{\Delta V} = \frac{N_i}{1 + \underline{k}} \; \frac{1 + \underline{k}}{\phi \; \Delta t \; u} = \frac{N_i}{\phi \; \Delta t \; u} \; . \tag{2-4}$$

In Eq. (2-4), Δt is the time interval between the beginning
and end of the elution of the zone at a carrier-gas velocity
u as measured under the conditions at the column outlet, ΔV
being the corresponding volume. The quantity Δt can be writ-
ten as $\Delta\underline{L}/u_i = \Delta\underline{L} \; (1 + \underline{k})/u$, and from the theoretical plate
theory [25] it follows that $\Delta\underline{L} = 4(H\underline{L})^{1/2}$, where H is the height
equivalent to a theoretical plate. In addition, ϕ can be ex-
pressed as $\varepsilon\pi\rho^2$, where ε is the total porosity of the column
packing and ρ is the inner radius of the empty column. Hence,
Eq. (2-4) can be rewritten to read

$$c_i' = \frac{N_i}{4\varepsilon\pi\rho^2 (H\underline{L})^{1/2}(1 + \underline{k})} \; . \tag{2-5}$$

This equation shows quite explicitly how c_i' depends, at a given
N_i, on the basic geometrical and chromatographical characteristics
of the column. The form of the equation can further be modi-
fied by expressing $\underline{L}$ and $\Delta\underline{L}$ in terms of the retention volume
V_R. The volume flow rate of the carrier gas (column effluent),
dV/dt, is ϕu, and (dV/dt) $\Delta t = \Delta V_R$. According to the plate-
height theory, the relation $\Delta V_R = 4V_R/\underline{N}^{1/2}$ applies, where $\underline{N}$ is
the number of plates, so that

$$c_i' = \frac{\underline{N}^{1/2}}{4V_R} \; N_i . \tag{2-6}$$

II. ZONE-MAXIMUM CONCENTRATION

The instantaneous solute concentration in the column effluent is a function of time; let us denote this concentration
$c_i(t)$. According to the mean-value theorem, the mean solute
concentration can be expressed as

$$c_i' = \frac{1}{\Delta t} \int_{t_1}^{t_2} c_i(t)\ dt \qquad\qquad (2-7)$$

which, on combining with Eq. (2-4), gives

$$\int_{t_1}^{t_2} c_i(t)\ dt = \frac{N_i}{dV/dt}\ . \qquad\qquad (2-8)$$

From the analytical viewpoint, the instantaneous solute
concentration in the maximum of the zone is of interest, as
there is a simple relationship between this concentration and
the total solute amount in the zone. The profile of the solute concentration in the column effluent, as measured at the
column outlet under constant conditions (carrier-gas flow rate,
temperature, pressure), approaches very closely a Gaussian distribution in linear nonideal chromatography. Hence, it is possible to deduce easily the relationship between the solute concentration in the zone maximum and the total amount of solute
chromatographed by considering the properties of the Gaussian
curve. For a Gaussian curve drawn in a coordinate system with
the abscissa b and ordinate h, it follows from the basic Gaus-
sian-curve theory that

$$\int_{b_1}^{b_2} h(b)\ db = h_m \sigma_b (2\pi)^{1/2}$$

where $h(b)$ denotes a value of the Gaussian function of an

independent variable b, h_m is the value of this function at the curve maximum, and σ_b is the standard deviation (expressed in the units of the property b) of the curve. The record of a chromatographic peak is more or less a corollary to a Gaussian curve in which h_m is identical with the peak height and b and σ_b are given by $(db/dt)t$ and $(db/dt)\sigma_t$, where db/dt is the recorder chart speed, t is time, and σ_t is the standard deviation expressed in time units. Hence,

$$\int_{b_1}^{b_2} h(b) \; db = \frac{db}{dt} \int_{t_1}^{t_2} h(t) \; dt,$$

and the above relation between $\int_{b_1}^{b_2} h(b) \; db$ and h_m can be rewritten as

$$\int_{t_1}^{t_2} h(t) \; dt = h_m \sigma_t (2\pi)^{1/2}.$$

Further, in the chromatographic peak, the quantities $h(t)$ and h_m are proportional to the corresponding instantaneous concentration $c_i(t)$ and the peak-maximum concentration c_i^*, respectively, the proportionality constant being the same in both cases (provided that the peak has been recorded within a linear range of the detecting and recording systems). It also holds, to a good approximation, that $\sigma_t = \Delta t/4$ (cf. [26]), so that we have

$$c_i^* = \frac{4}{(2\pi)^{1/2}} \frac{1}{\Delta t} \int_{t_1}^{t_2} c_i(t) \; dt. \tag{2-9}$$

A comparison of this equation with Eqs. (2-7) and (2-6) immediately gives

$$c_i^* = \frac{4 c_i'}{(2\pi)^{1/2}} = \frac{N^{1/2}}{(2\pi)^{1/2} V_R} N_i \tag{2-10}$$

where the factor $4/(2\pi)^{1/2}$ has a value of about 1.6. Hence,
the dependence of c_i^* on the basic parameters of the chromatographic column is essentially the same as the dependence of c_i'
on these parameters [cf. Eq. (2-5)], the only difference being
that $c_i^* = 1.6\ c_i'$.

Equations (2-5), (2-6), and (2-10) represent the relationship between the mean or peak-maximum concentration, the total
solute amount in the zone, and the properties of the chromatographic system. The detector senses the solute concentration
in the column effluent, giving a response which is, in a certain manner, related to the solute amount N_i. Although the
function of the detector can easily be imagined as independent
of the chromatographic column, the on-line combination of the
column and detector yields some new qualities typical of quantitative analysis by gas chromatography. For instance, as far
as the function of the detector alone is considered, the integral $\int_{t_1}^{t_2} c_i(t)\ dt$ is the only unequivocal measure of the absolute solute amount N_i. The fact that c_i^* is also related unambiguously to N_i is due to the mechanism of the chromatographic
process, i.e., due to the column. In nonlinear chromatography,
relations (2-5), (2-6), and (2-10) are inapplicable. Also, V_R
can be a significant parameter in GC quantitation; increasing
V_R will decrease c_i^* at a given value of N_i, thus decreasing the
sensitivity and reliability of analysis. Hence, in spite of
the fact that it is the detector that serves as the analyzer
proper in quantitative gas chromatography, the conception of
this discipline cannot be complete without considering the
properties of the column.

Chapter 3

DETECTION

I. DEFINITIONS OF FUNDAMENTAL TERMS

The following treatment often uses the terms analytical property, signal, and response. An analytical property, denoted by a, is, generally, any property of mass which is in a precisely defined relation to the quantity and quality of the substance to be determined [27]. A signal, denoted by S, is the instantaneous action of mass displaying a given analytical property on a sensing element which is able to respond to the analytical property. A response, denoted by R, is the instantaneous reaction of detector to a signal.

The analytical property of a mixture of the gaseous component (i) and carrier gas (o) and the analytical property of pure carrier gas are denoted a_{io} and a_o, respectively.

Accordingly, the signal of the mixture of component and carrier
gas and the signal of carrier gas alone are written S_{io} and
S_o; the respective responses are R_{io} and R_o. The net signal
produced by a component present in the column effluent is defined by $S_{io} - S_o$, and is called S_i. The net response recorded
as a deflection of the recorder pen from the baseline is $R_{io} - R_o$, and is denoted R_i.

II. CLASSIFICATION OF DETECTORS

In gas chromatography there are currently three modes of
detector classification. Classification of detectors according to the time dependence of their response allows distinction
between cumulative (or integral) and differential detectors.
The integral detectors signal the total amount of a substance
which has passed through the device from the startup to a given
moment, whereas the differential detector indicates the instantaneous quantity of substance detected present at a given moment in the sensing element. Another classification is nondestructive and destructive detectors [28], i.e., those detectors
where the process of detection does not involve irreversible
changes of the substance detected and those where the substance
is subjected to irreversible changes. A third classification
distinguishes between concentration- and mass-sensitive detectors. The problems of this classification have been dealt with
by Halász [29]. Concentration-sensitive detectors respond to
the concentration of a substance chromatographed in the column
effluent rather than to the total quantity of that substance
present in the sensing element or to the rate of introduction
of mass into the sensor. Mass-sensitive detectors respond to
the total amount of the substance detected in the sensing element and/or to the rate of introduction of the component detected into the sensing element. In context with the third

classification, the term concentration means the ratio, in
moles, of pure solute vapor to solute-vapor/carrier-gas mixture, rather than the solute amount per unit volume.

From the analytical viewpoint, it is appropriate to consider combinations of the above classifications, i.e., to
classify detectors as, e.g., differential/mass-sensitive/
destructive (flame-ionization detectors), differential/concen-
tration-sensitive/nondestructive (katharometer), etc. There
exist in general eight possible theoretical combinations. The
analytical significance of distinguishing between integral and
differential detectors is obvious from the graphs in Fig. 1.

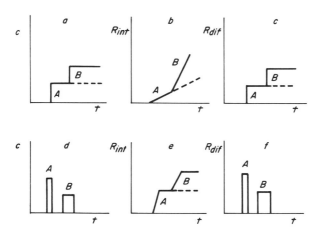

FIG. 1. Schematic representation of the dependence of
detector response R, on the corresponding time-dependence of
component concentration in column effluent, c; (a) time dependence of component concentration in column effluent in frontal
chromatography; (b) and (c) corresponding time dependences of
response of integral and differential detector; (d) time dependence of component concentration in column effluent in elution
chromatography; (e) and (f) corresponding response dependences
of integral and differential detectors.

Diagrams (a) and (d) illustrate the time dependence of concentration c of the components A and B in the effluent, and are
reminiscent of frontal and elution chromatograms. Plots (b)
and (c), as well as (e) and (f), show the type of record which
would be obtained in either case by using integral or differential detection. These diagrams show that, in frontal chromatography with integral detection, it is the slope of the respective
line which is a quantitative parameter of the chromatogram
[viz., (b)], while a quantitative parameter in elution (also in
displacement) chromatography with the same type of detection is
the height of the respective step in the record [viz., (e)].
With differential detectors, the shape of the record is similar
to the course of the time-dependent concentration profile, dis-
torted, however, because of the specific features of detection.
Chromatographic records obtained with differential and integral
detectors stand in the same relation as a direct and integrated
(by using an integrator) differential record. Bearing in mind
this simple relationship, we shall discuss only differential
detection and elution chromatography.

The classification of detectors as nondestructive and destructive is related to other analytical aspects. The signal
for nondestructive detectors is caused by the presence of the
substance analyzed and is produced as long as the material remains in the sensing element. It does not depend on the rate
of entry of the material into the sensing element. In destructive detectors, the signal is produced by some quantity involved
in the process of alteration of the active substance into an
inactive one. The products from some such process do not produce any signal. In this case, the signal is given by the rate
of entry of active mass into the sensing element. The consequences of this dependence are diagrammatically illustrated in
Fig. 2. In these diagrams, dV/dt is the volume flow rate of
column effluent through the sensing element, S^D and S^N are,

respectively, the signals detected with a destructive and a
nondestructive detector, and $\int S^D \, dt$ and $\int S^N \, dt$ are the time
integrals of the respective signals.   The amount of mass present, or its rate of entry, is, as a rule, proportional to the
response, so that diagrams (a) and (b) illustrate the flow dependence of the height of the chromatographic peak, and diagrams (c) and (d) show the flow dependence of the peak area.

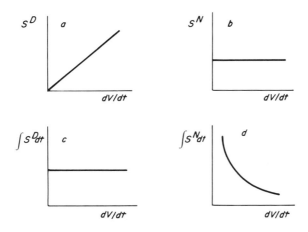

FIG. 2.   Dependence of signal S and of time integral of
signal, $\int S \, dt$, on volume flow of column effluent through the
sensing element, dV/dt, in detection with destructive detectors (a and c) and in detection with nondestructive detectors
(b and d).

The above types of flow dependence are still customarily
used as criteria for distinguishing between concentration and
mass detectors (cf. [29]).   In other words, the terms nondestructive and destructive are interchanged with the terms

concentration and mass. A more thorough analysis, however,
shows that this mingling is not justified, and that the classification of detectors as concentration- and mass-sensitive
detectors is related to the pressure dependence of the signal.
Here there are two processes conceivable, i.e., where pressure
changes occur at, respectively, (1) constant composition of
the effluent, and (2) constant amount of the component chromatographed in the effluent. In the first case, a change in
pressure does not involve any change in the relative concentration of the component chromatographed; the amount of mass in
the sensing element, however, will vary. The signal produced
by the substance chromatographed will therefore be proportional
to the pressure with mass-sensitive detectors, whereas with
concentration-sensitive detectors it will remain independent
of pressure. The same can be said, of course, of the time
integral of the signal. This state of affairs is illustrated
diagrammatically in Fig. 3. S^C and S^M denote, respectively,
the signals detected with a concentration and a mass detector,
and P is the overall pressure in the sensor.

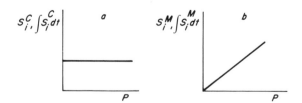

FIG. 3. Pressure dependence of signal S and of its time
integral, $\int S\ dt$, in detection with concentration-sensitive
(a) and mass-sensitive detector (b), assuming constant composition of column effluent.

In the second case, the total mass of the substance chromatographed (e.g., the mass determined by the amount injected)
is obviously independent of pressure. However, the concentration of the substance is, under these circumstances, inversely
proportional to the pressure, as at a given volume of the sensor the number of moles in it of the effluent is proportional
to the pressure. Hence, it immediately follows that the pressure dependence of the signal of the component i will follow
the course indicated in Fig. 4.   The conclusions arrived at for
1 and 2 above hold both for nondestructive and destructive detectors.   (The term effluent is employed for the mixture of
carrier gas and solute vapor, whether the gases are actually
leaving a column or are merely present in the detector sensor.)

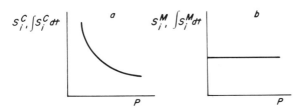

FIG. 4.   Pressure dependence of signal S and of time integral of signal, $\int S\,dt$, in detection with concentration-sensitive detectors (a) and mass-sensitive detectors (b), assuming
constant total quantity of component in column effluent.

III.   PERFORMANCE CHARACTERISTICS OF MODEL TYPES OF DETECTOR

The foregoing discussion can serve to establish criteria
for the evaluation of the basic characteristics of detectors.
These criteria are listed in Table 1.

TABLE 1

Survey of Criteria for Evaluation of Typical
Detector Characteristics

Detector type		Typical characteristics		Note[a]
Nondestructive (N)		$\dfrac{dS}{d(dV/dt)} = 0$	$\dfrac{d \int S\ dt}{d(dV/dt)} = -\dfrac{\int S\ dt}{dV/dt}$	a
Destructive (D)		$\dfrac{dS}{d(dV/dt)} = \text{const.}$	$\dfrac{d \int S\ dt}{d(dV/dt)} = 0$	b
Mass sensitive (M)	(1)	$\dfrac{dS_i}{dP} = \text{const.}$	$\dfrac{d \int S_i\ dt}{dP} = \text{const.}$	b
	(2)	$\dfrac{dS_i}{dP} = 0$	$\dfrac{d \int S_i\ dt}{dP} = 0$	b
Concentration sensitive (C)	(1)	$\dfrac{dS_i}{dP} = 0$	$\dfrac{d \int S_i\ dt}{dP} = 0$	b
	(2)	$\dfrac{dS_i}{dP} = -\dfrac{S_i}{P}$	$\dfrac{d \int S_i\ dt}{dP} = -\dfrac{\int S_i\ dt}{P}$	b

[a](a) P, T = const.; (b) dV/dt, T = const.

The characteristics listed in Table 1 of course refer to
idealized models of detectors. In most cases, the signal-re-
sponse process derived from the basic principle of detection is
accompanied by a number of extra processes which can considerably modify the expected behavior of the detector. Therefore,
any real detector will display certain deviations from the
properties that might be assumed in light of these criteria.
The individual criteria can be combined according to the
type of detector under consideration. Thus, by taking into
account the concepts discussed in Chapter 2 and considering the
situation existing in normal elution chromatography, it is

possible to characterize the concentration-sensitive/nonde-
structive (CN), mass-sensitive/nondestructive (MN), and mass-
sensitive/destructive (MD) detectors directly in terms of the
basic parameters of GC quantitation, i.e., the peak-maximum
solute concentration c_i^*, total number of moles of solute N_i,
peak height h_{mi}, and peak area A_i. Along with Eqs. (2-8) and
(2-10), the following relationships serve as the basis for
the derivation of the respective criteria:

$$h_{mi} \sim R_{mi} \tag{3-1}$$

$$A_i \sim \frac{db}{dt} \int_{t_1}^{t_2} R_i \, dt \tag{3-2}$$

where R_{im} is the peak-maximum net detector response, db/dt is
the recorder chart speed, and t_1 and t_2 are the times of the
beginning and the end of elution of the chromatographic zone.

A. *Concentration-Sensitive/Nondestructive Detectors*

Some CN detectors give a response proportional to the
solute concentration expressed as the number of moles of solute per unit volume of the column effluent, i.e., $R_i \sim c_i$.
In this case,

$$h_{mi} \sim c_i^* = \frac{N^{1/2}}{(2\pi)^{1/2} V_R} N_i \tag{3-3}$$

and

$$A_i \sim \frac{db/dt}{dV/dt} N_i \tag{3-4}$$

The gas-density detector should display these properties.

Other CN detectors detectors give a response proportional to the mole (volume) fraction (y_i) of the solute component in the column effluent, i.e., $R_i \sim y_i$. In this case,

$$y_i = \frac{N_{im}}{\underline{N}} = \frac{N_{im}RT}{P\underline{V}} = \frac{c_{i-}RT}{P} \tag{3-5}$$

where N and V are the number of moles and volume of the column effluent, T and P are the absolute temperature and pressure of the latter, and $\underline{R}$ is the gas constant. We can write

$$h_{mi} \sim y_i^* = \frac{RTN^{1/2}}{P(2\pi)^{1/2}V_R}N_i \tag{3-6}$$

and

$$A_i \sim \frac{RT(db/dt)}{P(dV/dt)}N_i \tag{3-7}$$

where y_i^* is the peak-maximum solute mole fraction in the column effluent. The behavior of the katharometer obeys Eqs. (3-6) and (3-7).

B. *Mass-Sensitive/Nondestructive Detectors*

MN detectors respond to the absolute mass of the solute present in the column effluent contained within the inner space of the detector sensor, $\underline{V}$, i.e., $R_i \sim c_{i-}\underline{V}$. For this case,

$$h_{mi} \sim c_{i-}^*\underline{V} = \frac{VN^{1/2}}{(2\pi)^{1/2}V_R}N_i \tag{3-8}$$

and

$$A_i \sim \frac{\underline{V}(db/dt)}{dV/dt}N_i \quad . \tag{3-9}$$

The cross-section, argon, and electron-capture detectors are
examples of this type of detector.

C.   Mass-Sensitive/Destructive Detectors

For MD detectors the response is proportional to the
rate of solute-mass supply into the detector sensor, i.e.,
$R_i \sim dN_{\underline{im}}/dt$.   As

$$\frac{dN_{\underline{im}}}{dt} = \frac{N_{\underline{im}}(dV/dt)}{V} = c_i \frac{dV}{dt} \, , \qquad (3\text{-}10)$$

we have

$$h_{mi} \sim c_i^* \frac{dV}{dt} = \frac{(dV/dt)N^{1/2}}{(2\pi)^{1/2}V_R} N_i \qquad (3\text{-}11)$$

and

$$A_i \sim \frac{db}{dt} N_i \qquad (3\text{-}12)$$

where dV/dt is again the volume flow rate of the column
effluent entering the sensor.   A typical representative of
this kind of detector is the flame-ionization detector.

The characteristics derived for the various detectors are
summarized in Table 2.   The proportionality constants relating
h_{mi} and A_i to the right-hand sides of the individual equations
shown in Table 2 are

$$C_h = \frac{Kf_i N^{1/2}}{(2\pi)^{1/2}V_R} \qquad (3\text{-}13)$$

$$C_A = Kf_i \frac{db}{dt} \qquad (3\text{-}14)$$

TABLE 2

Characteristics of Various Types of Detectors[a]

| Type of Detector | Dependence of h_{mi} and A_i versus N_i on dV/dt, T, P, and $\underline{V}$ | |
	h_{mi}	A_i
CN $(R_i \sim c_i)$	$h_{mi} \sim N_i$	$A_i \sim \dfrac{N_i}{dV/dt}$
CN $(R_i \sim y_i)$	$h_{mi} \sim \dfrac{T}{P} N_i$	$A_i \sim \dfrac{T}{P(dV/dt)} N_i$
MN	$h_{mi} \sim \underline{V} N_i$	$A_i \sim \dfrac{\underline{V}}{dV/dt} N_i$
MD	$h_{mi} \sim \dfrac{dV}{dt} N_i$	$A_i \sim N_i$

[a]CN = concentration-sensitive/nondestructive; MN = mass-sensitive/nondestructive; MD = mass-sensitive/destructive.

where K is the apparatus constant and f_i stands for a mass-specific response factor; the problems of the latter are dis--cussed in detail in the next chapter. The compliance between the criteria quoted in Tables 1 and 2 is readily apparent.

Chapter 4

RELATIONS BETWEEN PEAK AREA AND THE AMOUNT OF SOLUTE COMPONENT
IN THE CHROMATOGRAPHIC BAND

I.   ANALYSIS OF THE SIGNAL AND RESPONSE-DETERMINING PARAMETERS

The conclusions drawn in the discussion on detectors indicate that the relation between the amount of the substance
chromatographed and some quantitative parameter of the respective chromatogram will differ for various types of detector
used.   It is necessary that this fact be accounted for when
formulating these relations.   In this section, relations will
be derived between the analytical property and the signal produced, relations between the signal and response, and relations
between the total amount of the substance chromatographed and
the area of the chromatographic peak.   A signal is the primary
action of the substance detected on the sensing element.   The
response is the primary reaction of the sensing element to the

signal. It is obvious that the primary response may play the
role of a signal for a secondary response. Several such steps
may occur in a chain starting with some analytical property and
finishing with the output data of the detector system. These
relations will be discussed for four types of detector: mass-
sensitive/nondestructive (MN), mass-sensitive/destructive (MD),
concentration-sensitive/nondestructive (CN), and a detector
type which could be named "quasi-concentration-sensitive/
destructive" (CD). The last of these detector types is, in
principle, an MD detector, but, since for this detector the
signal is proportional to the ratio of the rate of introduction
of mass into the sensing element and the rate of reaction-prod-
uct formation, it is independent of the rate of entry of the
material, and the detector behaves as if it were concentration
sensitive. The Scott microflame detector [30] can be placed
in this group.

All considerations in this chapter are based on the following presuppositions:

1. Regardless of what state of aggregation the sample
possesses when introduced into the column, it moves along only
in the mobile phase, so that each component of the mixture
chromatographed is in a gaseous state when entering the sensor.

2. The rate at which the response becomes steady and the
maximum speed of the recorder pen are considerably higher than
the actual change of signal with time.

Presupposition 1 implies a constant mole and, consequently,
also volume flow rate of the column effluent through the sensing
element under steady hydrodynamic conditions in the chromatographic system, regardless of whether the effluent involves a
component chromatographed or the passing stream is merely pure
carrier gas. When injecting small samples, i.e., when the

effects of change in the viscosity of the gaseous medium can be
neglected, the above presupposition is met, owing to the fact
that pressure or flow disturbances incidental to the injection
of sample proceed at a much higher speed than the chromatographic zone. Hence, a steady flow is restored at the time of
zone entry into the sensing element, regardless of the character of the source of carrier gas, i.e., whether the source
functions as a source of constant pressure or of constant flow.

Fulfillment of presupposition 2 guarantees that the chromatographic peak is a true picture of the time dependence of
the signal. The only distorting factors are those which have
their origin in the very mechanism of response. These factors
are, as a rule, determined by the concentration of the substance chromatographed in the column effluent (i.e., by the
sample size) and by the operating conditions of the detector.
Presupposition 2 is not always met; the consequences have been
discussed in detail [20-22].

For the situation in which the total amount of the substance chromatographed is constant (see Section II of Chapter
3), we may write for the individual types of detector

$$S^{CN} = K_s a \tag{4-1}$$

$$S^{CD} = K_s a \, \frac{dN/dt}{dN_p/dt} \tag{4-2}$$

$$S^{MN} = K_s aN \tag{4-3}$$

$$S^{MD} = K_s a \, \frac{dN}{dt} \tag{4-4}$$

where K_s is a proportionality constant, N the total number of
moles of effluent in the sensing element of the detector, dN/dt
the molar rate of introduction of the effluent into the sensing
element, and dN_p/dt the molar rate of flow of reaction products

out from the sensing element. The analytical property a is to
be looked upon as an implicit function of the concentration of
the substance sensed by the detector. In all cases, the response will be presumed to be linearly proportional to the
signal, so that the relation

$$R = K_R S \qquad\qquad (4\text{-}5)$$

holds true. K_R is again a proportionality constant. The
product $K_S K_R$ is the overall apparatus constant, which will be
denoted K. For the net response,

$$R_i = K_R (S_{io} - S_o) = K'(a_{io} - a_o) \qquad\qquad (4\text{-}6)$$

where K' is the product of K and some pertinent parameter
characteristic of the type of detection under consideration
[compare relations (4-1)-(4-4)].

The relationship between the instantaneous amount of the
substance chromatographed in the column effluent and the corresponding net response of the detector, R_i, is determined by
the detection principle, which is independent of the nature of
the sorption system, and in general can be written as

$$R_i = R_i\left(D, \ T, \ \frac{dV}{dt}, \ \frac{dV_\alpha}{dt}, \ P, \ a_i, \ a_o, \ a_\alpha, \ y_i\right) \qquad\qquad (4\text{-}7)$$

where, of the symbols not yet defined, D is a factor involving
detector design parameters, T the temperature at which the
sensing-element body is being kept, dV_α/dt the volume flow rate
of the additional gas, a_α the analytical property of the additional gas, and y_i the mole fraction of the substance chromatographed in the column effluent. Relation (4-7) holds generally
for any detector type.

II.   LINEARITY OF RESPONSE

Relation (4-7) may serve as a basis for establishing three conditions for linearity of the relationship between the instantaneous amount of the substance chromatographed in the column effluent and the corresponding net detector response.

1.   The parameters D, T, dV/dt, dV_α/dt, P, a_i, a_o, and a_α are constant, and, at the same time, it is true that

$$a_{io} = a_i y_i + a_o y_o = (a_i - a_o)y_i + a_o \qquad (4\text{-}8)$$

where y_o is the mole fraction of the carrier gas in the column effluent ($y_o = 1 - y_i$). Relation (4-8) expresses a linear additivity of the analytical properties of the vapor of the component chromatographed and that of the carrier gas according to the composition of the mixture, which, with respect to the previously stated suppositions yields

$$R_i = K'(a_{io} - a_o) = K'(a_i - a_o)y_i \; . \qquad (4\text{-}9)$$

Equation (4-9) implies a linear response over the whole concentration range, which in practice only rarely occurs.

2.   In several cases, the analytical properties a_i and a_o are not additive, i.e., relation (4-8) is not valid. In these circumstances, with the other parameters constant (except for y_i), the response cannot be expected theoretically to be linear over any concentration range. In practice, the situation is more favorable, since, within narrow concentration limits, the nonlinear course can be taken as approximately linear. Such a situation, for a concentration range which is of interest from the view point of chromatography, is shown in Fig. 5. The dashed line illustrates a hypothetical linearly additive course; the full line represents the real course; and the dotted line represents an approximate linear course over a concentration range of about 0-0.1. It is obvious that this approximation

of the linearity of the a_{io} versus y_i plot cannot be taken as
an approximation of the linear additivity of the analytical
properties a_i and a_o.

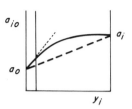

FIG. 5. Dependence of analytical property of the component-vapor/carrier-gas mixture, a_{io}, on component concentration, y_i, in the medium flowing through the sensing element.

The dependence of a_{io} on composition can generally be
expressed by

$$a_{io} = a_{io}(a_i, a_o, y_i). \qquad (4\text{-}10)$$

The right-hand side can be expanded into a McLaurin series, i.e.,

$$a_{io} = a_o + \frac{1}{1!}\frac{da_{io}}{dy_i}(0)y_i + \frac{1}{2!}\frac{d^2a_{io}}{dy_i^2}(0)y_i^2 + \cdots \qquad (4\text{-}11)$$

For the chromatographically significant concentration range of
the substance chromatographed in the column effluent, it is
sufficient to consider only the first two terms of the series,
i.e.,

$$a_{io} = a_o + \frac{da_{io}}{dy_i}(0)y_i \qquad (4\text{-}12)$$

which is a mathematical relation for the above-mentioned linear
approximation. From relation (4-12), it follows that

$$R_i = K' \frac{da_{io}}{dy_i} (0) y_i \qquad (4-13)$$

which is the equation of an asymptotic tangent line passing
through the origin. It can be easily seen that the expressions
$da_{io}(0)/dy_i$ and $a_i - a_o$ are simply related by

$$\frac{da_{io}}{dy_i} (0) = K_{io}(a_i - a_o) \qquad (4-14)$$

where K_{io} is a constant determined for the given type of detection by the properties of the components i and o. For linearly
additive a_i and a_o, K_{io} is obviously unity. Unless explicitly
stated otherwise, the general expression (4-13) will be used in
subsequent considerations. For the sake of simplicity, the
symbol (0) will be omitted.

3. The nonlinearity resulting from the virtual detection
principle can be overcome by a proper detector design and by an
appropriate choice of the working parameters. The performance
characteristics of the detector D can be modified so as to be
such a function of concentration y_i that it will bias off the
nonlinear influence of the concentration dependence of a_{io} (cf.
the use of restricting resistors in argon detectors [31, 32]).

III. RELATIONS BETWEEN THE INSTANTANEOUS AMOUNT OF THE SUBSTANCE CHROMATOGRAPHED IN THE SENSING ELEMENT AND THE DETECTOR RESPONSE

When using the perfect-gas state equation, then N and
dN/dt can be expressed as PV/RT and (dV/dt)P/RT. Relations
(4-1)-(4-4) can be then written, taking into consideration Eqs.
(4-5), (4-6), and (4-13) or (4-14) to read

$$R_i^{CN} = K \frac{da_{io}}{dy_i} y_i \qquad (4-15)$$

$$R_i^{CD} = K \frac{dV/dt}{dV_p/dt} \frac{da_{io}}{dy_i} y_i \tag{4-16}$$

$$R_i^{MN} = K \frac{PV}{RT} \frac{da_{io}}{dy_i} y_i \tag{4-17}$$

$$R_i^{MD} = K \frac{P}{RT} \frac{dV}{dt} \frac{da_{io}}{dy_i} y_i \tag{4-18}$$

where dV_p/dt is the volume rate of reaction products outflow, and $\underline{V}$ is the geometrical volume of the sensor space proper.

From Eq. (4-15), it obviously follows that the response is independent of the flow rate of effluent (carrier gas); at a given K and da_{io}/dy_i, the only variable is the concentration of the substance chromatographed, y_i. In the case described by (4-16), the state of affairs is practically the same, since dV_p/dt can be taken proportional to dV/dt according to the relation

$$\frac{dV_p}{dt} = \xi \frac{dV}{dt} \tag{4-19}$$

where ξ is a constant.

The response of mass-sensitive detectors is proportional not only to concentration y_i but also to the term $P/\underline{R}T$. The response of MN-type detectors is, in addition, proportional to the volume of the sensing element and independent of the flow rate. The response of the MD-type detectors is proportional to the flow rate and independent of the volume of the sensing element.

Upon introducing

$$\frac{dN_i}{dt} = \frac{dN}{dt} y_i = \frac{P}{\underline{R}T} \frac{dV}{dt} y_i \tag{4-20}$$

where dN_i/dt is the mole flow rate of only the chromatographed

substance in the effluent, relations (4-15)-(4-18) can be
rewritten as

$$R_i^{CN} = K \frac{\frac{RT}{-}}{(P(dV/dt)} \frac{da_{io}}{dy_i} \frac{dN_i}{dt} \qquad (4-21)$$

$$R_i^{CD} = \frac{K}{\xi} \frac{\frac{RT}{-}}{P(dV)/dt)} \frac{da_{io}}{dy_i} \frac{dN_i}{dt} \qquad (4-22)$$

$$R_i^{MN} = K \frac{\frac{V}{-}}{dV/dt} \frac{da_{io}}{dy_i} \frac{dN_i}{dt} \qquad (4-23)$$

$$R_i^{MD} = K \frac{da_{io}}{dy_i} \frac{dN_i}{dt} \qquad (4-24)$$

The ratio $(dN_i/dt)/(dV/dt)$ represents the concentration of solute eluted in the effluent, expressed in moles per unit volume.
Let this concentration be m_i. On introducing the corresponding
relation

$$\frac{dN_i/dt}{dV/dt} = m_i \qquad (4-25)$$

Eqs. (4-21)-(4-24) become

$$R_i^{CN} = K \frac{RT}{P} \frac{da_{io}}{dy_i} m_i \qquad (4-26)$$

$$R_i^{CD} = \frac{K}{\xi} \frac{RT}{P} \frac{da_{io}}{dy_i} m_i \qquad (4-27)$$

$$R_i^{MN} = KV \frac{da_{io}}{dy_i} m_i \qquad (4-28)$$

$$R_i^{MD} = K \frac{dV}{dt} \frac{da_{io}}{dy_i} m_i \qquad (4-29)$$

which show the direct dependence of the response on working
conditions.

IV.   RELATIONS BETWEEN THE TOTAL AMOUNT OF THE SUBSTANCE
CHROMATOGRAPHED PASSED THROUGH THE SENSING ELEMENT AND
THE TIME INTEGRAL OF THE DETECTOR RESPONSE

For analytical purposes, it is an advantage to relate the
total amount of a separated component of the mixture chromatographed, N_i, and the corresponding quantitative parameter of
the respective chromatographic peak.   Such relations can be obtained by integrating Eqs. (4-21)-(4-24) with respect to time
over the interval between the beginning and end of the elution
band, t_1 and t_2.   Thus, relations of the type

$$\int_{t_1}^{t_2} R_i \ dt = K' \ \frac{da_{io}}{dy_i} \ N_i \tag{4-30}$$

are obtained where

$$N_i = \int_{t_1}^{t_2} \frac{dN_i}{dt} \ dt \tag{4-31}$$

The instantaneous deflection of the recorder pen, h_i, is proportional to the net response:

$$h_i = \beta R_i \tag{4-32}$$

where β is an apparatus (recorder) constant.   The area of the
chromatographic peak, A_i, is given by

$$A_i = \frac{db}{dt} \int_{t_1}^{t_2} h_i \ dt \tag{4-33}$$

where db/dt is the recorder chart speed.   Combining Eqs. (4-33)
and (4-32),

$$A_i = \beta \ \frac{db}{dt} \int_{t_1}^{t_2} R_i \ dt \tag{4-34}$$

which, with respect to Eq. (4-30), leads to the following
equations:

$$A_i^{CN} = K\beta \frac{db}{dt} \frac{\frac{RT}{P(dV/dt)}}{} \frac{da_{io}}{dy_i} N_i \qquad (4-35)$$

$$A_i^{CD} = \frac{K}{\xi} \beta \frac{db}{dt} \frac{\frac{RT}{P(dV/dt)}}{} \frac{da_{io}}{dy_i} N_i \qquad (4-36)$$

$$A_i^{MN} = K\beta \frac{db}{dt} \frac{V}{dV/dt} \frac{da_{io}}{dy_i} N_i \qquad (4-37)$$

$$A_i^{MD} = K\beta \frac{db}{dt} \frac{da_{io}}{dy_i} N_i \qquad (4-38)$$

Relations (4-33), (4-34), and all those derived by virtue
of these two refer to literally the peak area as recorded on
the recorder strip chart. However, the concepts concerning the
analytical significance of the time integral of the detector
response apply independently of whether the latter is expressed
as a real peak area or another way. Hence, within this context
it is appropriate to conceive the term peak area in a broader
sense, viz., as a quantity proportional to the time integral of
the detector response.

When employing integrators for determining the peak area,
the detector response is integrated directly on the time basis,
so that the "peak area" obtained in this manner is not related
to the recorder chart drive. Usually, the response is converted
to the rate of some property proportional to the instantaneous
value of the former, which is further integrated within the time
limits t_1 and t_2. Provided that the instantaneous response is
proportional to the number of arbitrary counts per unit time,
$d\eta/dt$, the following equation holds:

$$\left(\frac{d\eta}{dt}\right)_i = \kappa R_i \tag{4-39}$$

where κ is a proportionality constant, and

$$\int_{t_1}^{t_2} \left(\frac{d\eta}{dt}\right)_i dt = \eta_i = \kappa \int_{t_1}^{t_2} R_i \, dt \tag{4-40}$$

where η_i is the overall number of counts, referred to component i. Comparison of Eq. (4-40) with Eq. (4-34) gives

$$\eta_i = \frac{\kappa}{\beta(db/dt)} A_i \tag{4-41}$$

Hence, there is a simple relationship between η_i and A_i.

Except for the disk integrators in which the rotation of the disk is synchronized with the recorder chart drive, the quantity η_i is independent of the latter.

Let us recall that T and P in the relations for both R_i and A_i stand for the temperature and pressure in the detector sensor when it is being entered by a given amount of the substance to be detected, with dV/dt and $\underline{V}$ denoting the volume flow rate (as measured in the sensor) of the effluent and the geometrical volume of the sensor space, respectively. Hence, the relations can be referred to the situation in which a certain amount of substance i has substituted, at a constant temperature T, such amount of the carrier gas in the sensor that the pressure P within the latter remains unchanged, to produce a net response R_i.

V. THE ROLE OF AN AUXILIARY GAS

In some cases, an additional independent stream of gas is introduced into the sensing element along with the column effluent. This stream can be either mixed with the effluent before it enters the sensing element, or admitted to the sensing

element via a separate inlet. Thus, in Scott's detector or the
flame-ionization detector, introduction of a combustible gas is
necessary to produce the flame. In the argon detector and the
electron-capture detector, the so called scavenger stream is
used to reduce the effective volume of the sensing element.

From the analytical viewpoint, it is important to introduce the auxiliary gas at the column outlet so that the composition of the additional stream will be constant and independent of changes in the composition of the column effluent at a
steady state. The effects of the auxiliary gas differ with the
type of detector.

Let α be the auxiliary gas, $a_{i o \alpha}$ the analytical property
of the mixture of component chromatographed, carrier gas, and
auxiliary gas, $a_{o \alpha}$ the analytical property of the mixture of
carrier and auxiliary gas, and y_i^* the mole fraction of the component chromatographed in the sensing element, after mixing the
column effluent and the additional gas. Then, it is possible
to write

$$a_{i o \alpha} = \frac{da_{i o \alpha}}{dy_i^*} y_i^* + a_{o \alpha} \tag{4-42}$$

From Eq. (4-14), we obtain

$$a_{i o \alpha} = K_{i o \alpha}(a_i y_i^* + a_o y_o^* + a_\alpha y_\alpha^*) \tag{4-43}$$

where $K_{i o \alpha}$ is the respective material constant, y_o^* and y_α^* are
the mole fractions of the carrier and auxiliary gas in the ‣
mixture of the added gas and column effluent. Obviously

$$y_o^* = 1 - y_i^* - y_\alpha^* \tag{4-44}$$

Combining Eqs. (4-43) and (4-44), and bearing in mind that y_α^*
is constant,

$$a_{i o \alpha} = K_{i o \alpha}(a_i - a_o)y_i^* + a_{o \alpha} \tag{4-45}$$

and

$$\frac{da_{io\alpha}}{dy_i^*} = K_{io\alpha} \, (a_i - a_o) = \frac{K_{io\alpha}}{K_{io}} \frac{da_{io}}{dy_i} \tag{4-46}$$

It is easily shown that

$$y_i^* = y_i \, \frac{dN/dt}{(dN/dt) + (dN_\alpha/dt)} \tag{4-47}$$

where dN/dt is the flow rate in mole units of column effluent and dN_α/dt is the flow rate in mole units of the added gas. If the added gas has the same composition as the carrier gas, then obviously $K_{io\alpha} = K_{io}$.

For the individual detector types, and with regard to Eqs. (4-15)-(4-18), we can write

$$R_i^{*CN} = K \, \frac{da_{io\alpha}}{dy_i^*} \, y_i^* \tag{4-48}$$

$$R_i^{*CD} = K \, \frac{(dV/dt) + (dV_\alpha/dt)}{(dV_p/dt) + (dV_{\alpha p}/dt)} \, \frac{da_{io\alpha}}{dy_i^*} \, y_i^* \tag{4-49}$$

$$R_i^{*MN} = K \, \frac{PV}{RT} \, \frac{da_{io\alpha}}{dy_i^*} \, y_i^* \tag{4-50}$$

$$R_i^{*MD} = K \, \frac{P}{RT} \left(\frac{dV}{dt} + \frac{dV_{\alpha p}}{dt} \right) \frac{da_{io\alpha}}{dy_i^*} \, y_i^* \tag{4-51}$$

where, in addition to symbols already used, R^* is the net detector response when auxiliary gas is used, dV_α/dt is the volume flow rate of the auxiliary gas, and $dV_{\alpha p}/dt$ is the volume flow rate of the reaction products of the latter. If the composition of the auxiliary gas does not change, then $dV_{\alpha p}/dt = dV_\alpha/dt$. Equations (4-48)-(4-51) can be treated in a way similar to Eqs. (4-15)-(4-18), i.e., they can be transformed to correspond in form to relations (4-21)-(4-24), (4-26)-(4-29), and (4-35)-(4-38). Substituting Eqs. (4-46) and (4-47) in (4-48)-(4-51), and using volume flow ratios instead of the mole flow rate ratios, we obtain

$$R_i^{*CN} = K \frac{K_{io\alpha}}{K_{io}} \frac{dV/dt}{(dV/dt) + (dV_\alpha/dt)} \frac{da_{io}}{dy_i} y_i \qquad (4-52)$$

$$R_i^{*CD} = \frac{K}{\xi^*} \frac{K_{io\alpha}}{K_{io}} \frac{dV/dt}{(dV_p/dt) + (dV_{\alpha p}/dt)} \frac{da_{io}}{dy_i} y_i \qquad (4-53)$$

$$R_i^{*MN} = K \frac{K_{io\alpha}}{K_{io}} \frac{PV}{RT} \frac{dV/dt}{(dV/dt) + (dV_\alpha/dt)} \frac{da_{io}}{dy_i} y_i \qquad (4-54)$$

$$R_i^{*MD} = K \frac{K_{io\alpha}}{K_{io}} \frac{P}{RT} \frac{dV}{dt} \frac{da_{io}}{dy_i} y_i \qquad (4-55)$$

where ξ^* is the ratio $[(dN/dt) + (dN_\alpha/dt)]/[(dN_p/dt) + (dN_{\alpha p}/dt)]$ Proceeding in the same manner as with the relations describing the cases without auxiliary gas, we obtain the relationships between the peak area (here A*) and the total number of moles of the substance chromatographed, N_i:

$$A_i^{*CN} = K^* \frac{RT}{P} \frac{1}{(dV/dt) + (dV_\alpha/dt)} \frac{da_{io}}{dy_i} N_i \qquad (4-56)$$

$$A_1^{*CD} = \frac{K^*}{\xi^*} \frac{RT}{P} \frac{1}{(dV/dt) + (dV_{\alpha p}/dt)} \frac{da_{io}}{dy_i} N_i \qquad (4-57)$$

$$A_i^{*MN} = K^*V \frac{1}{(dV/dt) + (dV_\alpha/dt)} \frac{da_{io}}{dy_i} N_i \qquad (4-58)$$

$$A_i^{*MD} = K\beta \frac{db}{dt} \frac{da_{io}}{dy_i} N_i = A_i^{MD} \qquad (4-59)$$

In Eqs. (4-56)(4-58), K^* is given by

$$K^* = K \frac{K_{io\alpha}}{K_{io}} \frac{db}{dt} \beta \qquad\qquad (4-60)$$

These relations show that, except for MD detectors, the stream
of the added gas decreases detection sensitivity. It is also
significant that the application of auxiliary gas results in
changes of the flow characteristics of CN, CD, and MN type detectors. At a high rate of flow of the additional gas, when
$(dV/dt) \ll (dV_\alpha/dt)$, the response R_i^* is obviously proportional
(with the above-mentioned detector types) to the flow rate dV/dt
(dV_α/dt and $dV_{\alpha p}/dt$ are constant), while the area A_i^* is practically independent of the flow rate. The above detectors be-
have as if they were related to the family of MD detectors,
i.e., in a way exactly opposite to their original versions
without any additional gas stream. This is a logical consequence of rapid scavenging of the column effluent from the sensing-element space, which leads to effects similar to those
brought about by destruction of the substance detected.

VI.   MOLAR AND RELATIVE MOLAR RESPONSE; CORRECTION FACTORS

     The relations between the rate of introduction of a sub-
stance chromatographed into the sensing element and the net
response of the detector as well as those between the total
amount of the substance and the peak area serve as a basis for
determining the factors required to correct the quantitative
parameters of the chromatogram. The definition of these factors, as based on the above relations, unambiguously determines
the units in which the analytical results should be expressed.
     A general quantitative characteristic of detection speci-
ficity is the molar response MR_i, defined by

$$MR_i = \frac{dR_i}{d(dN_i/dt)} \tag{4-61}$$

Relations (4-21)-(4-24) show that the molar response also involves a parameter characteristic of the detector type given and an apparatus constant, besides the parameter da_{io}/dy_i, which is specific for the given substance. As long as chromatographic experiment is performed under constant conditions, the nonspecific part behaves the same for any component of the chromatographed mixture, and it is appropriate to eliminate it. In order to achieve this, the relative molar response, RMR_i, has been introduced [18]. The relative molar response is the molar response of substance i related to the molar response of some appropriate reference compound r which is chromatographed under the same conditions. Accordingly,

$$RMR_{ir} = \frac{MR_i}{MR_r} = \frac{da_{io}/dy_i}{da_{ro}/dy_r} \tag{4-62}$$

In addition to eliminating the nonspecific quantities, this ratio can also help lower or eliminate an incidental dependence of the specific factor da_{io}/dy_i on the operating conditions (T, P, dV/dt). Using (4-62), it is possible to relate the rate of solute introduction into the sensing element to the net response by

$$\frac{dN_i}{dt} = \frac{1}{MR_r} \frac{R_i}{RMR_{ir}} \tag{4-63}$$

Relation (4-63) has the same form for all detector types, irrespective of the use of an additional gas stream.

The relative molar response can also be formulated using Eqs. (4-35)-(4-38), or (4-56)-(4-59). It is easy to recognize that

$$RMR_{ir} = \frac{da_{io}/dy_i}{da_{ro}/dy_r} = \frac{A_i/N_i}{A_r/N_r} \qquad (4\text{-}64)$$

whence, with Eq. (4-63),

$$N_i = \frac{1}{\beta(db/dt)\ MR_r}\ \frac{A_i}{RMR_{ir}} \qquad (4\text{-}65)$$

The factors $1/MR_r$ and $1/\beta(db/dt)MR_r$ are, under given conditions, the same for any component chromatographed.

The correction factors can be deduced from Eqs. (4-63) and (4-65). The factor required to convert the quantitative parameters of the chromatogram to mole units, i.e., f_i^N, is given by

$$f_i^N = \frac{1}{RMR_{ir}} \qquad (4\text{-}66)$$

If the results are to be expressed in weight units, then the correction factor must be defined another way. The relation can be easily derived by substituting for N_i in Eqs. (4-63) and (4-65)

$$N_i = \frac{W_i}{M_i} \qquad (4\text{-}67)$$

where W_i and M_i are the weight and molecular weight of the component chromatographed. Then

$$\frac{dW_i}{dt} = \frac{1}{MR_r}\ \frac{R_i M_i}{RMR_{ir}} \qquad (4\text{-}68)$$

and

$$W_i = \frac{1}{\beta(db/dt)\ MR_r}\ \frac{A_i M_i}{RMR_{ir}} \qquad (4\text{-}69)$$

so that the factor which converts the quantitative parameters of the chromatogram to weight units, f_i^W, can be defined by

$$f_i^W = \frac{M_i}{RMR_{ir}} \qquad (4-70)$$

Taking into account Eq. (4-67), relations (4-66) and (4-70) can be rewritten as

$$f_i^N = \frac{K_{ro}(a_r - a_o)}{K_{io}(a_i - a_o)} \qquad (4-71)$$

$$f_i^W = \frac{M_i K_{ro}(a_r - a_o)}{K_{io}(a_i - a_o)} \qquad (4-72)$$

Assuming linear additivity of a_i, a_r, and a_o, i.e., $K_{io} = K_{ro}$ = 1, Eqs. (4-71) and (4-72) can be simplified accordingly.

With regard to the relations used for deriving f_i^N and f_i^W, it is obviously possible to use the factors both for correcting the instantaneous net response produced, e.g., by admitting at a constant rate substance i from some calibrating device into the sensing element, and for correcting peak areas.

In a recent paper [33], attention was paid to the role of the background response brought about by stationary-phase bleed-ing. It was shown that the presence of the stationary-phase vapor in the carrier gas leads generally to a defined decrease in the net response. This decrease, expressed as the ratio of the actual (decreased) net response to that which would be mea-sured in the absence of the stationary-phase vapor, is given by the factor $1 - y_b[(da_{bo}/dy_b)/(da_{io}/dy_i)]$, where b stands for the stationary-phase vapor, and the meaning of the other sym-bols is the same as before. For the relative molar response, it follows that

$$RMR_{ir} = \frac{(da_{io}/dy_i) - y_b(da_{bo}/dy_b)}{(da_{ro}/dy_r) - y_b(da_{bo}/dy_b)} \qquad (4-73)$$

The decrease of the net response cannot be eliminated by any method of compensation for the background response (dual-column operation, electrical compensation). This phenomenon is there-fore very important in quantitative analysis by programmed-temperature gas chromatography.

VII.   ANALYTICAL SIGNIFICANCE OF UNCORRECTED QUANTITATIVE
       PARAMETERS OF THE CHROMATOGRAM

We must often decide whether the uncorrected peak areas on a chromatogram approach more closely mole or weight data. This question can again be answered on the basis of examination of Eqs. (4-35)-(4-38) or (4-56)-(4-59). Assuming linear addi-tivity of analytical properties a_o and a_i [cf. Eqs. (4-8) and (4-9)], we may write Eq. (4-65) as

$$A_i = \frac{\beta(db/dt)\,MR_r}{a_r - a_o}\,(a_i - a_o)N_i \qquad (4\text{-}74)$$

If $a_i \ll a_o$ and, at the same time, $a_r \ll a_o$, then both a_i and a_r can be neglected relative to a_o, so that

$$A_i = \beta\,\frac{db}{dt}\,MR_r\,N_i \qquad (4\text{-}75)$$

Equation (4-75) shows that under the above conditions, the un-corrected area is proportional to the number of moles of the substance chromatographed. If, on the contrary, $a_i \gg a_o$ and $a_r \gg a_o$, then a_o can be neglected and Eq. (4-74) becomes

$$A_i = \frac{\beta(db/dt)\,MR_r}{a_r}\,a_i N_i \qquad (4\text{-}76)$$

Equation (4-76) implies that $f_i^N = 1/(a_i/a_r)$. In several cases, especially for the members of a homologous series, the analyti-cal property a is proportional to the molecular weight of the respective substance, M, i.e., we may write $a_i = C_{(i)}M_i$ and $a_r = C_{(r)}M_r$, where the C's are proportionality constants.

If again $a_i \gg a_o$ and $a_r \gg a_o$, then

$$A_i = \frac{\beta(db/dt)\, MR_r\, C_{(i)}}{C_{(r)}\, M_r}\, W_i \qquad (4\text{-}77)$$

which testifies for direct proportionality of uncorrected areas to the weights of components. It must be emphasized that Eqs. (4-74)-(4-77) are true only if linear additivity of analytical properties exists. The material constants K_{io} and K_{ro} can change the situation to a considerable extent.

Chapter 5

PREDICTION OF THE RELATIVE MOLAR RESPONSE

The goal of this chapter is to show the possibility of theoretically estimating the relative molar response of the more important detector types. In order to characterize clearly the analytical property, signal, and response, each detector type is briefly described in terms of the pertinent detection principle. Because the number of symbols used in this chapter is considerable, several of the symbols used to denote other quantities in the remaining sections will here be given other specific meaning.

I.   KATHAROMETER [34-40]

The column effluent passes through a cell, which comprises an element heated by a constant input of electrical energy.

The katharometer body has a considerably higher heat capacity
than the heated element and the gaseous medium in the cell, and
is thermostated. In a steady state, there is a constant temperature drop between the heated element and the inner surface of
the cell. The heat generated is transported to the element
environment. If a chromatographic fraction enters the cell, a
corresponding change of the capability to transfer heat results
and, consequently, the temperature of the heated element is
changed. Temperature changes cause corresponding changes in
the electrical resistance of the element and are biased off by
means of a bridge circuit.

In this case, the analytical property is the capability
for heat dissipation into the environment, the signal is the
actual heat dissipation, and the net response is given by the
difference in the temperature of the heated element on the
passage of a zone of a given instantaneous composition (T_{fio})
and that at passage of pure carrier gas (T_{fo}). Thus,

$$R_i = T_{fio} - T_{fo} \tag{5-1}$$

The heat flux supplied is given off from the element by
conduction through the gas to the inner surface of the cell
(q_{cd}), by convection (q_{cv}), by radiation (q_r), and by dissipation through the electrical connections (q_p). At a steady
state, it is obvious that

$$q = q_{cd} + q_{cv} + q_r + q_p \tag{5-2}$$

For a cylindrical cell of length L and radius r_c, with a coaxi-
ally positioned heated wire of radius r_f, there holds

$$q_{cd} = \frac{2\pi Lk}{\ln(r_c/r_f)} (T_f - T_c) \tag{5-3}$$

where T_f and T_c are the temperatures of the wire and of the
inner wall of the cell, respectively, and k is the heat

conductivity coefficient of the medium surrounding the wire.
For q_{cv},

$$q_{cv} = \frac{dN}{dt} C_p (T' - T_c) \qquad (5-4)$$

where dN/dt is the molar flow rate through the cell, T' is the
mean temperature of the gas leaving the cell, and C_p is the
mean molar heat capacity of the gas in the cell (assuming that
the temperature of the entering gas is T_c). For T', we may
write

$$T' = \frac{2}{r_c^2 - r_f^2} \int_{r_f}^{r_c} r\, T(r)\, dr \qquad (5-5)$$

which, for a cylindrically shaped cell, gives

$$T' = \frac{q_{cd}}{2\pi Lk} \left[\frac{r_f^2}{r_c^2 - r_f^2} \ln \frac{r_f}{r_c} + \frac{1}{2} \right] + T_c \qquad (5-6)$$

From relations (5-3) and (5-6), we obtain

$$\frac{T' - T_c}{T_f - T_c} = \frac{1}{2 \ln(r_c/r_f)} - \frac{r_f^2}{r_c^2 - r_f^2} \qquad (5-7)$$

On neglecting q_r and q_p with respect to q_{cv} and q_{cd}, and
$r_f^2/(r_c^2 - r_f^2)$ with respect to $1/[2 \ln(r_c/r_f)]$, then, on employing (5-3), (5-4), and (5-7),

$$T_f - T_c = q \frac{\ln(r_c/r_f)}{2\pi Lk + (dN/dt)C_p/2} \qquad (5-8)$$

from which we immediately obtain for the net detector response

$$R_i = T_{fio} - T_{fo} = q \ln \frac{r_c}{r_f} \left[\frac{1}{2\pi Lk_{io} + \frac{1}{2} \frac{dN}{dt} C_{pio}} - \frac{1}{2\pi Lk_o + \frac{1}{2} \frac{dN}{dt} C_{po}} \right].$$

$$\qquad (5-9)$$

Expanding Eq. (5-9) into a McLaurin series and taking the first
two members of each of its components,

$$R_i = \frac{-q \ln(r_c/r_f)}{\left(2\pi L k_o + \frac{1}{2}\frac{dN}{dt}C_{po}\right)^2}\left[2\pi L(k_{io} - k_o) + \frac{dN}{dt}\frac{C_{pio} - C_{po}}{2}\right] \qquad (5\text{-}10)$$

The compound heat conductivity k_{io} can be expressed by Wassiljewa's equation [41]

$$k_{io} = \frac{k_i}{1 + A_{oi}(y_o/y_i)} + \frac{k_o}{1 + A_{io}(y_i/y_o)} \qquad (5\text{-}11)$$

where, for a given pair of gases i and o, and under given conditions, A_{io} and A_{oi} are constants. For a narrow range of
solute (i) concentrations in the carrier gas, Eq. (5-11) can
again be expanded into a McLaurin series to yield

$$k_{io} = k_o + \left(\frac{k_i}{A_{oi}} - k_o A_{io}\right)y_i \qquad (5\text{-}12)$$

The constants, A_{oi} and A_{io} are defined by

$$A_{oi} = \left(\frac{\sigma_i + \sigma_o}{2\sigma_i}\right)^2 \left(\frac{M_i + M_o}{2M_o}\right)^{\frac{1}{2}} \qquad (5\text{-}13)$$

$$A_{io} = \left(\frac{\sigma_i + \sigma_o}{2\sigma_o}\right)^2 \left(\frac{M_i + M_o}{2M_i}\right)^{\frac{1}{2}} \qquad (5\text{-}14)$$

where the σ's are kinetic molecular effective collision diameters and the M's are molecular weights. Combining (5-10) and
(5-12), and assuming that C_{pi} and C_{po} are linearly additive,

$$R_i = \frac{-q \ln(r_c/r_f)}{[2\pi L k_o + (dN/dt)C_{po}/2]^2}\left[2\pi L\left(\frac{k_i}{A_{oi}} - k_o A_{io}\right) + \frac{dN}{dt}\frac{C_{pi} - C_{po}}{2}\right]y_i \qquad (5\text{-}15)$$

whence the relative molar response is given by

$$
RMR_{ir} = \frac{2\pi L \left(\dfrac{k_i}{A_{oi}} - k_o A_{io} \right) + \dfrac{dN}{dt} \dfrac{C_{pi} - C_{po}}{2}}{2\pi L \left(\dfrac{k_r}{A_{or}} - k_o A_{ro} \right) + \dfrac{dN}{dt} \dfrac{C_{pr} - C_{po}}{2}}
\tag{5-16}
$$

Equation (5-15) indicates that the katharometer is a CN-type
detector. Theoretically derived RMR values agree well with
corresponding experimental data [18, 42-44] as long as the
carrier gas is either hydrogen or helium. In other cases [37],
there was an evident discrepancy between the calculated and
measured data. This discrepancy probably arises because, over
the concentration range of solute i in the carrier gas normal
in chromatography, Eq. (5-11) is not valid.

II. MARTIN'S GAS-DENSITY BALANCE [45-48]

When pure carrier gas flows through one vertical tube,
while the column effluent passes through another, then if the
diameters of both tubes are sufficiently large to eliminate any
dynamic effects of flow, the difference between the pressures
at the inlets of the tubes is given by the difference in the
hydrostatic pressures in the tubes, Δp. The hydrostatic pressure is proportional to the product of the gas-column height
and gas density. If the tubes are connected near their inlets
by a piece of capillary tube of length ℓ and radius r, then
the difference in the hydrostatic pressures is proportional to
the flow of gas through the capillary.

The device can be set up in such a way that only the pure
carrier gas of viscosity η_o can flow through the capillary
tube. The mean flow rate $\overline{dV/dt}$ is given by the Poiseuill
equation

$$\overline{dV/dt} = \frac{\pi r^4}{8\eta_o \ell} \Delta p \tag{5-17}$$

The analytical property is the capability for altering the hydrostatic pressure of the gas column, the signal is the hydrostatic pressure proper, and the response is the flow of gas through the capillary tube. A change in the difference of the hydrostatic pressures, $\Delta p_{io} - \Delta p_o$, caused by presence of the solute in the column effluent is given by

$$\Delta p_{io} - \Delta p_o = g(d_{io}h_{io} - d_o h_o) = \frac{Pg}{RT} (M_{io}h_{io} - M_o h_o) \tag{5-18}$$

where d is the density, h is the tube height, g is gravitational acceleration, M is the molecular weight, and P and T are the overall pressure and temperature in the tubes. Assuming a linear additivity of M_i and M_o, the net response is

$$R_i = (\overline{dV/dt})_{io} - (\overline{dV/dt})_o$$

$$= \frac{\pi r^4 Pg}{8\eta_o \ell RT} \left[(M_i - M_o)y_i h_{io} - M_o (h_{io} - h_o) \right] \tag{5-19}$$

If h_{io} equals h_o, then $(\overline{dV/dt})_o$ is zero. The relative molar response is given by the relation

$$RMR_{ir} = \frac{M_i - M_o}{M_r - M_o} \tag{5-20}$$

Equation (5-19) shows that the gas-density balance is an MN detector (responding to solute mass per unit volume). The validity of Eq. (5-20) has been demonstrated in an indirect way by using it to measure molecular weights [49].

III. SCOTT'S MICROFLAME DETECTOR [30]

The column effluent is mixed with a constant stream of hydrogen, and the mixture enters a jet where a small flame burns at the orifice. The surrounding space is supplied with air. The temperature of the hydrogen flame with the carrier gas alone is the background response of the detector. In the presence of a component chromatographed in the column effluent, the temperature of the flame alters correspondingly. The temperature is measured by a thermocouple. The analytical property is the capability to bring about a change of flame temperature, the signal is the heat evolved on combustion, and the net response is the difference in flame temperature in the presence and absence of a solute component in the column effluent.

For the combustion of pure hydrogen,

$$aH_2 + \frac{a}{2} O_2 + pI = aH_2O + pI + a \, \Delta H_{H_2} \tag{5-21}$$

where I is an inert component. If the pure carrier gas effluent, E_o, is introduced into the flame, then the reaction (5-21) is accompanied by

$$bE_o + n_o O_2 + m_o I = r_o P_o + m_o I + b \, \Delta H_o \tag{5-22}$$

where P denotes the products of combustion. If the effluent contains a chromatographic fraction (E_{io}), then the following reaction takes place instead of (5-22):

$$bE_{io} + n_{io} O_2 + m_{io} I = r_{io} P_{io} + m_{io} I + b \, \Delta H_{io} \tag{5-23}$$

In Eqs. (5-21)-(5-23), ΔH is the corresponding molar enthalpy of combustion under the conditions prevailing in the flame.

If the combustion is considered to be an adiabatic process, then, with regard to Eq. (5-21), the following heat balances exist for the processes (5-22) and (5-23):

$$a\, \Delta H_{\underline{H}_2} + b\, \Delta H_o + \int_{T_c}^{T_o} [aC_{p\underline{H}_2\underline{O}} + (m_o + p)C_{pI}$$

$$+ r_o C_{pP_o}]\, dT = 0 \qquad\qquad (5\text{-}24)$$

$$a\, \Delta H_{\underline{H}_2} + b\, \Delta H_{io} + \int_{T_c}^{T_{io}} [aC_{p\underline{H}_2\underline{O}} + (m_{io} + p)C_{pI}$$

$$+ r_{io} C_{pP_{io}}]\, dT = 0 \qquad\qquad (5\text{-}25)$$

where T_c, T_o, and T_{io} are, respectively, the column temperature, the flame temperature for pure carrier gas, and the flame temperature in the presence of a chromatographic fraction containing a component i. The symbols C_p represent the corresponding molar heat capacities. By introducing the mean molar heat capacities C_p', integrating Eqs. (5-24) and (5-25) over the limits given, and subtracting the integrated (5-24) from the integrated (5-25), we obtain an expression for the net detector response:

$$R_i = T_{io} - T_o = \frac{-(b\, \Delta H_{io} - a \Delta H_{\underline{H}_2})}{aC_{p\underline{H}_2\underline{O}}' + (m_{io} + p)C_{pI}' + r_{io} C_{pP_{io}}'}$$

$$- \frac{-(b\, \Delta H_o - a \Delta H_{\underline{H}_2})}{aC_{p\underline{H}_2\underline{O}}' + (m_o + p)C_{pI}' + r_o C_{pP_o}'} \qquad (5\text{-}26)$$

This equation can be considerably simplified by assuming that

$$C_{pP_{io}}' \doteq C_{pP_o}' \qquad\qquad (5\text{-}27)$$

$$m_{io} - m_o \ll m_o + p \qquad\qquad (5\text{-}28)$$

$$r_{io} - r_o \ll r_o \qquad\qquad (5\text{-}29)$$

Then, the net response can be expressed by

$$R_i = - \frac{b(\Delta H_{io} - \Delta H_o)}{ac'_{pH_2O} + (m_o + p)C'_{pI} + r_o C'_{pP_o}} \tag{5-30}$$

Assuming linear additivity of ΔH_i and ΔH_o, we obtain

$$R_i = - \frac{b(\Delta H_i - \Delta H_o)}{ac'_{pH_2O} + (m_o + p)C'_{pI} + r_o C'_{pP_o}} y_i \tag{5-31}$$

and the relative molar response is given by

$$RMR_{ir} = \frac{\Delta H_i - \Delta H_o}{\Delta H_r - \Delta H_o} \tag{5-32}$$

For an incombustible carrier gas, ΔH_o is obviously zero.

It can be concluded from the description of the response
mechanism and from Eq. (5-31) that the Scott detector is a CD
detector.

Equation (5-32) conforms to the measured data of Henderson
and Knox [50]. The discrepancies between the measured and calculated data, found by Bullock [51] and Primavesi [52], are
likely caused by an irregular response mechanism brought about
by an excessively high rate of introduction of combustible mass
(large sample size) into the sensing element.

IV. FLAME-IONIZATION DETECTOR [53-56]

The basic design of a flame-ionization detector is similar
to Scott's detector. The detection principle, however, is
based on a measurement of the electrical conductivity of the
flame. The mechanism of the response origin has been explained
by Sternberg et al. [57] as chemi-ionization. If pure hydrogen
is burning in air, then there exists in the burning flame a

certain proportion of atomic oxygen and hydrogen, together with oxygenated fragments, e.g., $\underline{OH}$ and $\underline{O_2H}$. On considering monoatomic oxygen and hydrogen only, the state of affairs can be expressed by

$$a(\underline{H}_2 + \frac{1}{2} \underline{O}_2 + pI) = 2ay\underline{H} + ay\underline{O} + apI + (1 - y)\underline{H_2O} \qquad (5\text{-}33)$$

If a substance chromatographed enters the flame, e.g., a hydrocarbon $\underline{C}_n\underline{H}_m$, it is first thermally cracked according to the following scheme:

$$NC\underline{_n}\underline{H}_m \rightarrow nN\underline{CH} + r\underline{R} \qquad (5\text{-}34)$$

where $\underline{R}$ denotes a moiety. The $\underline{CH}$ radicals react with oxygenated fragments or with oxygen atoms at a diffusion-controlled rate. If ay $\gg$ Nn, which holds for N $\ll$ a, then we may write, e.g.,

$$Nn\underline{CH} + Nn\underline{O} \rightarrow Nn\underline{CHO}* \qquad (5\text{-}35)$$

This reaction is highly exothermic, and the liberated reaction energy transforms the reaction products into excited species. The excitation energy cannot be "thermalized" at a sufficient rate, so that a part of this energy is consumed by the ionization process. The process can be illustrated by the scheme

$$Nn\underline{CHO}^* \rightarrow \alpha Nn\underline{CHO}^+ + \alpha Nne^- + (1 - \alpha)Nn\underline{CHO} \qquad (5\text{-}36)$$

where α is the degree (efficiency) of ionization. The analytical property is the capability of producing ions in the course of combustion, the signal is the actual formation of ions, and the net response is the ionization current.

From Eq. (5-36),

$$R_i = C\alpha n \frac{dN_i}{dt} \qquad (5\text{-}37)$$

where C is a proportionality constant. The net response (i.e., the ionization current) is therefore proportional not only to

the molar rate of entry of the substance chromatographed into
the sensing element, but also to the number of carbon atoms in
the molecule of the substance. Since atoms other than carbon
are also contributing, to a varying extent, to the ionization
current, while the contribution of carbon varies with its
function in the molecule, the concept of "effective carbon,"
C_{ef}, must be introduced. This quantity represents the contribution of a given atom to the ionization current, as expressed
in the units of that of the parafinic carbon atom. Thus, Eq.
(5-37) can be given a more general form, i.e.,

$$R_i = C\alpha (\Sigma \ C_{ef})_i \ \frac{dN_i}{dt} \qquad\qquad (5-38)$$

which, for the relative molar response, leads to

$$RMR_{ir} = \frac{(\Sigma \ C_{ef})_i}{(\Sigma \ C_{ef})_r} \qquad\qquad (5-39)$$

Equations (5-37) and (5-38) illustrate that the flame-
ionization detector is an MD detector.

Equation (5-39) is well obeyed for hydrocarbons, though
more accurate measurements have shown [58] that various isomeric
hydrocarbons, having the same number of carbon atoms in the
molecule, produce slightly different relative molar responses.
For compounds also containing atoms other than C and H in the
molecule, the departures of measured from calculated data are
so considerable [59] that the latter can only be used to roughly
estimate the results. From the viewpoint of analysis, it is of
significance that the RMR values for a homologous series are
proportional [55, 60, 61] to the number of carbon atoms within
wide limits. The significance of functional groups as sources
of discrepancies between measured and calculated data obviously
decreases as the chain length increases.

Recent papers on the flame-ionization detector [62-64] show that the net response of this detector is appreciably dependent on pressure. This phenomenon, which is actually at variance with the concept of the flame-ionization detector as an MD detector, is likely due to variations of the shape and pattern of the flame upon changing the pressure.

V. CROSS-SECTION IONIZATION DETECTOR [65-67]

Under normal conditions, gases are insulators. However, if exposed to radioactive radiation, they will be ionized and become electrically conductive. If the dimensions of the sensing element (ionization chamber) are small compared to the range of flight of the ionizing particles, i.e., if using a source sufficiently "hard" in conventional detectors, it can be expected that the concentration distribution of charged particles over the space of the sensing element is uniform, while the rate of ionization, expressed as the rate of generation of secondary electrons, dn/dt, in a steady state is proportional to the product of length ℓ over which the ionizing particles remain active, the number of moles per irradiated area, N^*, and the activity of the source, e^0. Thus, it can be written that

$$\frac{dn}{dt} = Qe^0N^*\ell \qquad\qquad (5\text{-}40)$$

where the proportionality constant Q represents the so called effective molar ionization cross section, characterizing the ionizability of a molecule, and the product $N^*\ell$ indicates the number of moles in the exposed volume. A particle ionizes a large number of molecules along its path, so that when a β emitter is being used, the number of secondary electrons in a given space will be considerably larger than the number of primary electrons.

If the ionized gas is exposed to an electric field so strong that all charged particles generated will be discharged at the electrodes sooner than they can recombine, then the detector operates in the range of saturated current, which is proportional to the rate of formation of the charged particles:

$$I = 1.602 \times 10^{-19} \, Q e^0 N^* \ell \tag{5-41}$$

The analytical property is the ionizability by the energy of primary particles of the radioactive source, the signal is the generation of ions, and the response is the ionization current.

If pure carrier gas flows through the sensing element,

$$I_o = 1.602 \times 10^{-19} \, Q_o e^0 N_o^* \ell \tag{5-42}$$

If a chromatographic fraction enters the sensing element, i.e., if certain number of the carrier-gas molecules are replaced by the molecules of substance i having an ionization cross section Q_i, the ionization current will be altered correspondingly. If the mean ionization cross section of a mixture of solute and the carrier gas leaving the column is denoted by Q_{io},

$$I_{io} = 1.062 \times 10^{-19} \, Q_{io} e^0 N_{io}^* \ell \tag{5-43}$$

where, obviously, $N_{io}^* = N_o^* = N^*$. The net response is given by the difference $I_{io} - I_o$, so that

$$R_i = 1.062 \times 10^{-19} \, e^0 N^* \ell (Q_{io} - Q_o) \tag{5-44}$$

If $N^* \ell$ is expressed by the equation of state, and linear additivity of the ionization cross sections Q_i and Q_o is assumed, then

$$R_i = 1.062 \times 10^{-19} \, e^0 \frac{PV}{RT} (Q_i - Q_o) y_i \tag{5-45}$$

whereupon the relative molar response is given by

$$RMR_{ir} = \frac{Q_i - Q_o}{Q_r - Q_o} \qquad\qquad (5-46)$$

Relation (5-45) indicates that cross-section ionization detectors belong to the family of MN detectors. The possibility of theoretically predicting the correction factors for the detection with a cross-section ionization detector has been verified for hydrocarbons by Boer [66].

VI. ELECTRON-CAPTURE DETECTOR [5, 68-71]

If the flowing carrier gas is exposed to irradiation by a soft β emitter (T, ^{63}Ni), then in close proximity to the emitter, in the so-called plasma, a certain concentration of electrons is established. Besides the primary electrons, secondary electrons exist in the plasma produced in the ionization process, which occurs as a result from collisions of primary electrons with the gas molecules present. Of course, positive ions are also produced in this reaction. Application of an electric field in the plasma results in the generation of a standing current. Because of the low energy of electrons (approaching the thermal energy) and the relatively low potential drop across the ionization chamber, the exposed space becomes an ideal medium for electron capture. This capture starts as soon as molecules of some solute possessing affinity toward electrons appear in this space. This may lead to direct formation of negative molecular ions, or the molecule may be destroyed to form neutral and negative fragments of the original molecule. Either reaction results in a decreased ionization current, due to the lower mobility of the negative molecular ions as compared to that of the electrons, which in addition enhances the probability of their perishing by recombination with the positive ions present.

The analytical property is the capability of decreasing the concentration of free electrons under the conditions described, the signal is the actual electron capture, and the response is the drop of background current.

Let us take a planar layer of the column effluent, bordered on one side by a parallel emitter surface acting as a cathode, and on the other side by a parallel anode. Let e^0 be the rate of direct electron generation on the cathode, and $(dn/dt)_\ell$ be the electron flux across the corresponding cross section at a distance ℓ from the cathode. If N^* is the number of moles per total cross-sectional area of the exposed gas, then

$$-d\left(\frac{dn}{dt}\right)_\ell = \varepsilon e^0 N^*\, d\ell \tag{5-47}$$

where ε is the molar extinction coefficient. Relation (5-47) is a corollary to the Lambert-Beer law. For pure carrier gas passing through the sensing element, and for the appearance of a chromatographic fraction in this gas,

$$-d\left(\frac{dn}{dt}\right)_{\ell o} = \varepsilon_o e^0 N_o^*\, d\ell \tag{5-48}$$

$$-d\left(\frac{dn}{dt}\right)_{\ell io} = \varepsilon_{io} e^0 N_{io}^*\, d\ell \tag{5-49}$$

which, after integrating and inclusion of the relation $N_{io}^* = N_o^* = N^*$, become

$$\left(\frac{dn}{dt}\right)_{\ell o} = e^0 \exp(-\varepsilon_o N^* \ell) \tag{5-50}$$

$$\left(\frac{dn}{dt}\right)_{\ell io} = e^0 \exp(-\varepsilon_{io} N^* \ell). \tag{5-51}$$

By expanding Eqs. (5-50) and (5-51) into a McLaurin series, and taking into account the first two terms of the series, we obtain

$$\left(\frac{dn}{dt}\right)_{\ell io} - \left(\frac{dn}{dt}\right)_{\ell o} = - e^0 N^* \ell (\varepsilon_{io} - \varepsilon_o) \qquad (5\text{-}52)$$

The product $N^* \ell$ is the number of moles present in the control space (sensing element), $\underline{V}$. Assuming linear additivity of ε_i and ε_o, and expressing $N^* \ell$ by the perfect-gas state equation,

$$\left(\frac{dn}{dt}\right)_{\ell io} - \left(\frac{dn}{dt}\right)_{\ell o} = - e^0 \frac{P\underline{V}}{RT} (\varepsilon_i - \varepsilon_o) y_i \qquad (5\text{-}53)$$

The difference $(dn/dt)_{\ell io} - (dn/dt)_{\ell o}$ is, under given conditions (sensing element geometry, voltage), proportional to the difference of the corresponding ionization currents, $I_{io} - I_o$, which represents the net detector response. Thus,

$$R_i = \frac{Ce^0 P\underline{V}}{RT} (\varepsilon_i - \varepsilon_o) y_i \qquad (5\text{-}54)$$

where C is a proportionality constant.

From Eq. (5-54), it follows that electron-capture detectors belong to the group of MN detectors. The relative molar response can be written as

$$RMR_{ir} = \frac{\varepsilon_i - \varepsilon_o}{\varepsilon_r - \varepsilon_o} \qquad (5\text{-}55)$$

VII.   ARGON IONIZATION DETECTOR [72-75]

The basic design of this detector is similar to that of
the cross-section ionization detector, but the use of noble
gas as the carrier and application of high accelerating voltage across the electrodes lead to a qualitatively different
mechanism of the response origin and, consequently, to different detector properties.

If a molecule of the carrier gas gains energy lower than
its ionization energy, the molecule can become excited. This
property is important for noble gases. The excitation energy
for polyatomic molecules is rapidly distributed among the
individual degrees of freedom, i.e., it becomes thermalized,
thus rendering the life of these excited molecules, and com-
parably their concentration, negligible. The life span of
excited noble-gas atoms extends to 10^{-4} sec. These atoms do
not themselves contribute to the conductivity of the exposed
medium. They can, however, ionize most organic molecules by
energy transfer.

As long as the strength of the electric field guarantees
a saturation ionization current, while the accelerated electrons do not acquire enough energy to excite the atoms of the
noble gas, the detector behaves as a cross-section detector.
Starting with a voltage E_e, the accelerated electrons gain
sufficient energy to bring about the excitation of the carrier-
gas molecules. This results in a sharp increase in concentration of the excited atoms, while the background current will
not increase. If molecules of a solute enter this medium, they
acquire the excitation energy and ionize, and the ionization
current increases.

The analytical property is the capability of becoming ionized by the excitation energy transferred from activated atoms
of the noble gas, the signal is the occurrence of ions, and

the net response is the change in the ionization current, I_{io} - I_o. The following relation has been derived [73, 75, 76] for this change:

$$I_{io} - I_o = \frac{\sigma_i v_i y_i (\alpha + \varphi I_o)}{\sigma_i v_i y_i (1 - \varphi) + \sigma_o v_o y_o}$$

(5-56)

where

$$\varphi = a \exp[b(E - E_e)]$$

(5-57)

In Eqs. (5-56) and (5-57), σ, v, and α are, respectively, the effective molecular kinetic diameter, the mean relative velocity of the molecules, and the rate of formation of excited argon atoms; E is the applied voltage, and a and b are constants. Under constant working conditions and at small concentrations of the solute eluted, i, both φ and α remain virtually constant, and the term $\sigma_i v_i y_i (1 - \varphi)$ in the denominator of Eq. (5-56) can be neglected with respect to $\sigma_o v_o y_o$. The relative molar response is then

$$RMR_{ir} = \frac{\sigma_i v_i}{\sigma_r v_r}$$

(5-58)

From the kinetic theory of gases,

$$\sigma = \left(\frac{6}{N\pi d}\right)^{1/3} M^{1/3}$$

(5-59)

$$v = [1.13(2RT)^{1/2}] \frac{1}{M^{1/2}}$$

(5-60)

where N is the Avogadro number, d is density, and M the molecular weight. Combining Eqs. (5-58), (5-59), and (5-60),

$$RMR_{ir} = \left(\frac{M_r}{M_i}\right)^{1/6} \tag{5-61}$$

From the form of Eq. (5-56), it is not readily seen in which detector group the argon ionization detector belongs. However, it is clear from the concept involved that this detector belongs to the MN group.

The possibility of predicting the correction factors for detection with an argon ionization detector has been verified by Lovelock [72] for a variety of compounds including n-aliphatic alcohols, n-aliphatic fatty acids and their methylesters, ketones, ethers, and aromatic hydrocarbons. The results of measurements agreed well with predictions except for aromatic hydrocarbons. Another paper [56] concerns the discrepancies between calculated and measured correction factors for alcohols. These difficulties can originate in other processes competing with the collision mechanism represented by Eq. (5-56), e.g., straightforward ionization proportional to the ionization cross section, which is not included in the equation above. The matter is further complicated by the fact that the processes occurring in the detector can acquire entirely different facets by merely changing the operating conditions, design features, and carrier-gas purity (cf. [23]).

Chapter 6

CONVENTIONAL TECHNIQUES

Because of a wide variety of problems which the analyst
is expected to attack, the choice of a proper analytical technique is a difficult task. The goal of this chapter is to fa-
cilitate this task by presenting a consistent survey of the
individual techniques of gas-chromatographic quantitative
analysis. In order to gain a suitable insight into the logic
of this very useful subject, each of the techniques is discussed on the basis of mathematical expressions. The derivations made stem from a common theoretical ground set forth
in Chapter 4. The discussion will encompass only those techniques which are concerned with the determination of the components of a sample under analysis. Those techniques which
have been developed to investigate detector performance are
omitted [77-79].

I. PROBLEM ANALYSIS

The procedure involved in quantitative analysis by gas chromatography includes the following steps:

1. Sampling and sample treatment before injecting the sample into a gas chromatograph

2. Sample injection, chromatography of the charge, and recording the chromatogram

3. Measuring quantitative parameters of the chromatographic record

4. Interpretation of the data obtained from measurements of the chromatographic record

The term "technique of quantitative analysis by gas chromatography" is understood as the procedure to be followed in carrying out these four operations. Each of these steps can present difficulties. The analyst should be well aware of possible difficulties in order to be able to cope with them, or, at least, to compensate for factors which adversely influence the precision and accuracy of analytical results. Thus, for example, the preparation of a representative sample can be a problem per se. Sometimes, prior to gas chromatography, it is necessary to homogenize the sample, to dispose of ballasts, or to concentrate the components of interest by some other physical or chemical method, e.g., filtration, centrifugation, distillation, extraction, etc. These operations can be accompanied by changes in sample composition, which sometimes are not easily defined. Another preliminary treatment often used prior to gas chromatography is the preparation of more volatile derivatives from compounds of low volatility. Esterification, etherification, acetylation, trifluoroacetylation, and silylation are the commoner reactions used. Sometimes, the products of chemical degradation are determined. Such steps, of course,

can involve the introduction of further errors, stemming from
the nonquantitative course of the reaction.

Sample injection can be impaired by part of sample escap-
ing around the syringe needle or, generally, by leaks in the
dosing device. If sample injection is accompanied by partial
fractionation, the loss by leakage does not conform in proportion to the composition of bulk sample.

There are several unfavorable situations which may occur
in the actual chromatography. A nonlinear sorption isotherm
(peak asymmetry) results in a nonlinear dependence of peak
height on the amount of sample injected [80], even when the
corresponding area dependence remains linear. A gross depar-
ture from linearity can bring about complete disappearance of
small peaks [81]. Some solutes may be irreversibly sorbed by
a given stationary phase, or can resist chromatography under
the conditions used (low volatility), which again disfigures
the results. Similar effect can be brought about by decomposition of a component in the injection port or in the column
[82]. Sometimes, an earlier deposit may be purged by introducing a new charge either from the injection port or from the
column (ghosting effect) [83]. Another disagreeable situation
arises when the concentration of some component in the column
effluent exceeds the linear range of detector response, even
when the total quantity of sample does not yet overload the
column. These complications result in distortion of peaks,
both in shape and area, and sometimes even in peak inversion
[84]. With rapidly eluted fractions, peak distortion may be
due to the time constant of the detection and recorder systems
or to low speed of the recorder pen [20-22].

If the separation is incomplete, especially in those situations where it is necessary to measure the area of a small
peak appearing on the shoulder of a large one, an auxiliary
base line must be drawn, which introduces doubt as to both the

peak height and area. An adverse state of affairs can be
encountered if the band of the solute to be determined is overlapped by the band of some other component which does not give
any signal in the detector. The presence of such interfering
substance, particularly when it is present in a large excess
over the substance to be determined, usually distorts the shape
and size of the peak of the latter in an undefinable way [85].

Major factors playing a role in the choice of the technique are the performance characteristics of the apparatus
available, the character of the detector, and the accessibility
of standard compounds of sufficient purity for calibration purposes and for sample dilution.

The above discussion shows that the choice of the technique to be used must be made for each particular problem to be
solved and with regard to the performance of available instrumentation. It is conceivable that the first goal is to obviate
any possible error by choosing conditions under which the sources
of such error are excluded rather than by seeking some special
technique. In any case, however, it is necessary to be aware of
all such sources of errors. Presentation of some universal procedure which would give precise and accurate results under any
circumstances is not feasible.

From Eqs. (4-35), (4-38), (4-56), and (4-59), and (4-65),
the relation between the total number of moles of component i in
the chromatographic band, N_i, and the area of the corresponding
peak, A_i, can be written in the form

$$A_i = C \ RMR_{ir} \ N_i \tag{6-1}$$

where C is a constant and RMR_{ir} is the relative molar response
(the molar response of compound i, relative to the molar
response of a reference compound r). For a given detector,
operating under constant conditions, C will be the same for

any charge and any component of the mixture chromatographed.
This constant, of course, includes a factor of detection sensitivity, or, in other words, the value of C is inversely proportional to detector-sensitivity attenuation. This feature
must be borne in mind when switching the attenuator during a
run or between individual runs. If the peaks are not recorded
at the same attenuation, or if the chromatogram of a standard
sample is recorded at some other attenuation than the chromatogram of the sample analyzed, it is necessary to multiply
each area by the appropriate attenuation factor.

II.   PRESENTATION OF CONCENTRATION

In practice, the usual task is to determine the concentration of component i in the sample. The quantity N_i can be expressed in several ways. The form of expressing the concentration depends on the choice of concentration units in which the
results are to be presented.

In this book, we shall consider four kinds of concentration
units: weight amount per unit volume, q, number of moles per
unit volume, m, weight fraction g, and mole fraction x. The
concentration in moles per unit volume will be termed "molarity,"
although in analytical chemistry this term is strictly reserved
for a particular case of the former, viz., the number of moles
per liter. As there is a simple relationship between the concentration expressed in molarity and in weight per unit volume,
the equations referring to the individual conventional techniques
will be expressed for calculation in units of molarity and weight
fraction only; in some cases, the mole fraction will also be considered. The special techniques of quantitative gas chromatography will be treated in terms of weight per unit volume.

These concentractions, as specified for component i, for
instance, are defined as follows:

(1)   Weight/volume: $q_i = \dfrac{W_i}{V_{(i)}} = \dfrac{M_i N_i}{V_{(i)}} = M_i m_i$ (6-2)

(2)   Molarity: $m_i = \dfrac{N_i}{V_{(i)}} = \dfrac{q_i}{M_i}$ (6-3)

(3)   Weight fraction: $g_i = \dfrac{W_i}{\sum_k W_j} = \dfrac{W_i}{W_{(i)}} = \dfrac{N_i M_i}{W_{(i)}}$ (6-4)

(4)   Mole fraction: $x_i = \dfrac{N_i}{\sum_k N_j} = \dfrac{W_i/M_i}{\sum_k (W_j/M_j)} = \dfrac{W_i/M_i}{W_{(i)}/M'_{(i)}}$

$$= \dfrac{g_i M'_{(i)}}{M_i} = \dfrac{q_i M'_{(i)}}{d_{(i)} M_i} = \dfrac{m_i M'_{(i)}}{d_{(i)}}$$
(6-5)

In Eqs. (6-2)-(6-5), k is the number of components in the mixture analyzed, W is weight, M is molecular weight, and V is volume. The primed symbols represent the mean values, those with subscripts without parentheses refer to pure substances, and those with subscripts in parentheses refer to the material containing the component in question. Capital letters W and V denote weights and volumes involved in the preparation of the sample for analysis; small letters w and v are used to denote the weight and volume of the sample which is injected into the chromatograph. Quantities referring to a substance used as the standard will carry the subscript s. The symbol r will again denote the reference compound used for expressing the relative molar response.

    For a standard,

$A_s = C \, RMR_{sr} \, N_s$ (6-6)

Assuming the working conditions to be constant, and combining
Eqs. (6-1) and (6-6),

$$N_i = \frac{RMR_{sr} A_i}{RMR_{ir} A_s} N_s \qquad (6-7)$$

Equation (6-7) is the basis of any technique of quantitative
analysis by gas chromatography.   Equations (6-1) and (6-7) can
be generally written as

$$A = C\ RMR\ mv_{(\ )} \qquad (6-8)$$

$$A = \frac{C\ RMR\ gw_{(\ )}}{M} \qquad (6-9)$$

where the blank parentheses are to stress that the v and w refer to the charge of the mixture to be analyzed for the content
of component i or s.   Equation (6-7) can be rearranged in a
similar manner.   The ratio RMR_{sr}/RMR_{ir}, which characterizes
those techniques where a standard substance is used other than
the substance to be determined, can be calculated either directly
from known RMR values or by analysis of a sample with known contents of the components i and s.   For the ratio RMR_{sr}/RMR_{ir}, the
following general relation holds true [cf. Eq. (4-64)]:

$$\frac{RMR_{sr}}{RMR_{ir}} = \frac{A_s N_i}{A_i N_s} \qquad (6-10)$$

It is obvious that, for experimental determination of the above
ratio, it is not necessary to know the absolute amount of the
mixture injected.   Only the ratio N_i/N_s need be known, and it
is identical to the corresponding mole-fraction ratio x_i/x_s or
molarity ratio m_i/m_s.

III. SURVEY OF WORKING TECHNIQUES

A. *Absolute-Calibration Technique*

The principle of this technique rests with separately injecting defined quantities of the sample analyzed and of a standard substance and subsequently comparing the areas under the
chromatographic peaks. Both injections must be made under identical conditions. Generally, any available sufficiently pure
substance which is amenable to chromatography under existing
conditions can be used as a standard. However, it is necessary
to know the relation between the relative molar response of the
standard and of the compounds to be determined [cf. relation
(6-7)]. When standardizing with some substance other than the
one of analytical interest, one cannot compare the chromatographic peak areas on the basis of their heights, except for
the particular case when the peak of the standard is eluted as
rapidly and spread to the same extent as the peak of the substance of interest. If the standard is identical with the compound to be determined, one may use the heights for the calculations. In such case, however, it is advisable that the mode of
injection and the amount injected be the same in either case.
The technique of absolute calibration can be employed in either
of the following ways.

1. *Direct-Comparison Method*

If using this method, the chromatograms obtained from injecting the sample analyzed and the standard substance are
interpreted directly on the basis of relation (6-7). By (6-8)
and (6-9), we obtain for the calculation of molarity

$$m_i = \frac{RMR_{sr} \, v_{(s)} A_i}{RMR_{ir} \, v_{(i)} A_s} \, m_s \qquad\qquad (6-11)$$

and for the calculation of the weight fraction

$$g_i = \frac{RMR_{sr} \; M_i w_{(s)} A_i}{RMR_{ir} \; M_s w_{(i)} A_s} g_s \tag{6-12}$$

Either relation shows that if molarities are to be considered,
it is necessary to measure the injected volumes, whereas in calculating the weight fractions, it is necessary to know the weight
of the sample injection. If measuring, in the latter case, volumes instead of weights, it is not possible to simply substitute
the quantities $w_{(\;)}$ by $v_{(\;)}$; the w's must be replaced by the
products $v_{(\;)} d_{(\;)}$, where d is density. It is obvious that, in
general, $d_{(i)}$ and $d_{(s)}$ are not identical.

2. Calibration-Curve Method

In order to employ this method, a plot of the amount of
the standard injected against the magnitude of the corresponding corrected quantitative parameter of the chromatogram must
first be constructed. It is of advantage to plot the product
$v_{(s)} m_s$ or $w_{(s)} g_s$ against A_s/RMR_{sr} and $A_s M_s/RMR_{sr}$, respectively.
These procedures are diagramatically shown in Fig. 6. In both
cases, the calibration graphs should be linear with a slope
equal to the inverse of the constant C and should, in both cases,
cross the origin of the coordinate system. According to the
calibration plots, the expressions A_i/RMR_{ir} and $A_i M_i/RMR_{ir}$ give
directly the number of moles or the weight of substance i, i.e.,
for a known $v_{(i)}$ or $w_{(i)}$ the molarity and the weight fraction of
the substance of interest in the sample. As long as analysis
and calibration are performed under identical conditions, the
calibration curve, constructed in the above manner, can be used
quite generally, i.e., it can be used in the analysis of any substance, independently of the substance used for calibration.
This generality is made possible by the fact that corrected

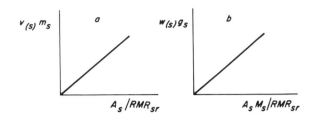

FIG. 6. Calibration graphs for the determination of
molarity (a) and weight fraction (b) by the technique of absolute calibration. ($v_{(s)}$, $w_{(s)}$, volume and weight of standard
injected; m_s, g_s, molarity and weight fraction of standard in
calibrating solution; A_s, peak area of standard; RMR_{sr}, relative molar response of standard; M_s, molecular weight of
standard.)

quantitative data of the chromatogram are plotted. This technique even permits construction of a calibration curve from
points obtained from different substances. If the calibration
is carried out with the pure substance to be determined, then
not only does correction become superfluous, but the calibration curve is valid both for areas and heights of the chromatographic peaks.

When constructing the calibration curve, either different
quantities (v, w) of the standard sample of the same concentration or the same amounts of samples differing in concentration
may be injected. The latter mode is preferred if peak heights
are to be used for the calculation. A detailed study of the
statistical aspects of this problem has been published [86].

The technique of absolute calibration is of advantage in
those cases where it is necessary to eliminate errors caused by
incomplete course of a reaction or by losses in sample adjust-
ment prior to injection into the instrument. Furthermore,
errors caused by high peak asymmetry, partial irreversible

sorption, partial decomposition, or by nonlinearity of the detector response are reduced. This is true only if the calibration curve is prepared using the pure substance of analytical
interest as the standard. If errors stemming from sample treatment prior to chromatography are to be eliminated, it is necessary that the calibration samples be treated in the same manner
as the actual sample.

If, for any reason, it is not feasible to add any extraneous compound to the sample under analysis, and if, at the same
time, the sample contains a component unamenable to chromatography or detection, or an unidentifiable portion, the method of
absolute calibration is the only possible means for quantitative
evaluation of the chromatogram. The method of absolute calibration suffers from the disadvantage that the absolute amount of
the substance injected into the chromatographic apparatus must
be known. If the analysis conditions cannot be kept identical
with those of calibration, the method of absolute calibration
cannot be used.

B.   *Internal-Standardization Technique*

It is obvious by (6-8) and (6-9) that the concentration of
the component to be determined, both in mole and weight units,
can be determined from only the ratio of the amounts of standard and sample injected, i.e., $v_s/v_{(i)}$ and $w_s/w_{(i)}$. In principle, this ratio can be determined prior to injection into the
instrument, so that the absolute amounts injected need not be
measured. This is the principle on which the technique of
internal standardization is based. In practice, we proceed so
that a given, defined amount of the sample under analysis $(V_{(i)},$
$W_{(i)})$ is mixed with a known amount of the standard (N_s, W_s) and
the resulting mixture is then injected into the chromatograph.

There are two variants to this method.

1. *Direct-Comparison Method*

First consider the determination of molarity. If we injected a volume $v_{(i)}$ of the original sample of molarity m_i, then the area of the corresponding chromatographic peak A_i would be given by (6-8). Upon adding the standard, the initial molarity m_i is changed (reduced) to m_i'; the molarity of the standard is m_s'. If $v_{(i)}'$ be the injected volume of a sample-standard mixture and A_i' and A_s' the corresponding peak areas, then

$$A_i' = C \ RMR_{ir} \ m_i' v_{(i)}' \tag{6-13}$$

$$A_s' = C \ RMR_{sr} \ m_s' v_{(i)}' \tag{6-14}$$

so that

$$\frac{A_i'}{A_s'} = \frac{RMR_{ir} \ m_i'}{RMR_{sr} \ m_s'} \tag{6-15}$$

The molarities m_i, m_i', and m_s' are

$$m_i = \frac{N_i}{V_{(i)}} \tag{6-16}$$

$$m_i' = \frac{N_i}{V_{(i)} + V_s} \tag{6-17}$$

$$m_s' = \frac{N_s}{V_{(i)} + V_s} \tag{6-18}$$

so that Eq. (6-15) can be rearranged to

$$m_i = \frac{N_s}{V_{(i)}} \ \frac{RMR_{sr}}{RMR_{ir}} \ \frac{A_i'}{A_s'} \tag{6-19}$$

The relation in terms of weight fractions can be derived in a
similar way, i.e.,

$$g_i = \frac{W_s}{W_{(i)}} \frac{RMR_{sr}}{RMR_{ir}} \frac{M_i A'_i}{M_s A'_s} \tag{6-20}$$

Both relations illustrate that the absolute amount of the sample injected need not be known; the ratios $N_s/V_{(i)}$ and $W_s/W_{(i)}$
are used instead. Theoretically, when using this technique, an
analysis can be carried out by injecting only a single charge
of a defined mixture of the sample and added standard, the necessary condition for a correct result is a linear relationship
between the areas of the component determined as well as that
of the standard and the total amount of either component in the
respective chromatographic band.

2. Calibration-Curve Method

This method is based on the preparation of a series of
known solutions varying in the concentration of the component
to be determined, i, to which known amounts of standard are
added. Chromatography of these solutions, followed by plotting
m_i or g_i against $N_s A'_i/V_{(i)} A'_s$ and $W_s A'_i/W_{(i)} A'_s$, yields the respective calibration curves. The procedure is illustrated in Fig.
7. The calibration curves thus obtained hold for the given
pair of compounds, i and s, and are independent of the ratios
$N_s/V_{(i)}$ or $W_s/W_{(i)}$. In either case, the calibration curves
should be straight lines with slopes equal to the terms
RMR_{sr}/RMR_{ir} (when using m_i in the calculations) or
$RMR_{sr} M_i/RMR_{ir} M_s$ (if using g_i) and should cross the origin of
the coordinate system. The analysis of an actual sample is
performed by following an analogous procedure; the concentrations m_i or g_i are determined from the corresponding values
of $N_s A'_i/V_{(i)} A'_s$ and $W_s A'_i/W_{(i)} A'_s$ according to the respective
calibration curves.

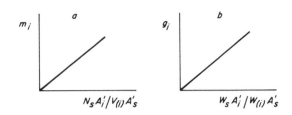

FIG. 7.   Calibration graphs for the determination of
molarity (a) and weight fraction (b) by the technique of internal standardization.   (N_s, W_s, mole number and weight of standard added to sample volume $V_{(i)}$ or weight $W_{(i)}$; A_i', A_s', peak
areas of component and standard recorded after injection of
sample/standard mixture; for other symbols see Fig. 6.)

The applicability of calibration graphs obtained for a
particular pair of compounds i and s can also be extended to
other pairs, i.e., to such cases where either another standard
is used, another substance is determined, or both.   For such
purposes, it is necessary to work with corrected quantitative
parameters of the chromatogram, i.e., to plot m_i or g_i against
$N_s RMR_{sr} A_i'/V_{(i)} RMR_{ir} A_s'$ and $W_s RMR_{sr} M_i A_i'/W_{(i)} RMR_{ir} M_s A_s'$.   The
slopes of the calibration graphs are obviously unity in both
cases.   In this case, the calibration graph can be theoretically constructed, as with the technique of absolute calibration, from points obtained quite generally from different pairs
of compounds i and s.   If, on the other hand, the same pair of
compounds i and s is used both for the calibration and the
actual analysis, and if the ratio $N_s/V_{(i)}$ or $W_s/W_{(i)}$ is kept
the same for all calibrating solutions as well as for the solution under analysis, then we can merely plot m_i or g_i against
the corresponding ratios A_i'/A_s'.   However, utilization of such
a calibration graph is restricted.   Of the three methods using
a calibration graph, the first, illustrated in Fig. 7, is the
most often used.

A main advantage of the internal-standard technique is the
ability to determine the component in question without knowing
the absolute amount of sample injected. A disadvantage is the
necessity of adding to the sample analyzed an extraneous admix-
ture which may become a source of difficulties. Thus, the
technique cannot be used if there is no free space in the chromatogram for the peak of the standard. Another major advantage
of the internal-standard technique is that both the component
to be determined and the standard are introduced into the instrument by a single injection. Under such conditions, it is
easily ensured that the bands of the component under analysis
and of the standard are eluted and recorded under like conditions. The requirement for identical working conditions during
both calibration and actual analysis, which fully applies to
the technique of absolute calibration, is reduced to a requirement for constant working conditions during one chromatographic
run for the technique of internal standardization. It is appropriate, from this viewpoint, to have the bands of standard and
the component of analytical interest appearing as near to each
other as possible. The sole limitation is that there be sufficient resolution of the bands. If, however, it is possible to
guarantee constancy of the operating conditions during chromatography, then the question of relative spacing of the peaks
of the component and the standard in the chromatogram will
have little bearing.

Of greater importance is the problem of the type and
amount of the standard added. The requirements for a suitable
standard can be summarized as follows: The standard must be a
compound which is well separated from all components of the

mixture analyzed under existing operating conditions.
The standard shall not react with any component of the sample,
nor should it influence the physical properties of the other
components, e.g., their volatility. The volatility of the standard present in a sample ready to be introduced into the chromatograph should approach the volatility of the component to be
assayed. Finally, a standard compound should, under conditions
chosen, yield a symmetrical chromatographic peak. The amount
of standard to be added ought to be comparable to the content
of the sample component to be determined, so that chromatography
of the component and the standard will take place under conditions corresponding to linear portions of the respective sorption isotherms, and the maximum concentration of either compound
in the column effluent will not exceed the linear range of detector response. If these requirements are satisfied, then,
once constructed, a calibration curve remains valid even if
the actual analysis is carried out under other conditions, unless such an alteration would result in diverging RMR values
[87].

C. *Standard-Addition Technique*

This technique is related to internal standardization.
With regard to the procedure adopted for sample preparation for
chromatography, the only difference is that no new component
but the component to be determined is added as the standard to
the sample. The fact that the standard added is not separated
by chromatography results in a different and more complicated
interpretation of the chromatogram. A major feature of all
types of this technique is that two injections are required,
the same as when using the technique of absolute calibration.
There are three types of the standard-addition technique.

1. *Method of Direct Measurement of the Sample Injected*

The first step is the chromatography of a known amount of
the original sample. If the calculation is in terms of molarity, then, for the area under the curve of the component determined, relation (6-8) is applicable. The next step is to add
a measured amount of the standard and to inject a measured
amount of this enriched sample. Both chromatograms are run
under the same conditions.

The standard added will, as previously, be denoted s, and
all quantities related to the standard will bear the subscript
s in spite of the standard and the component determined being
identical. In other words, the substance present in the original sample and the substance added (i) will be viewed as if
they were different components. According to this concept, the
concentration of the original component under analysis in the
enriched sample will be lower. If we denote the quantities relating to the enriched sample with a prime, then with regard to
relation (6-8), we can write

$$A'_i = C\ RMR_{ir}\ m'_i v'_{(i)} \qquad\qquad (6\text{-}21)$$

$$A'_s = C\ RMR_{sr}\ m'_s v'_{(i)} \qquad\qquad (6\text{-}22)$$

It is obvious that the areas A'_i and A'_s form parts of a single
area, which is obtained upon injecting the enriched sample.
Let this area be A'_{is}. For this area,

$$A'_{is} = A'_i + A'_s. \qquad\qquad (6\text{-}23)$$

For the ratio of the areas obtained by respectively injecting
the enriched and the original sample, i.e., A'_{is}/A_i, we obtain
by Eqs. (6-8), (6-21), and (6-22)

$$\frac{A'_{is}}{A_i} = \frac{v'_{(i)}}{v_{(i)}}\left(\frac{m'_i}{m_i} + \frac{m'_s}{m_i}\right) \qquad\qquad (6\text{-}24)$$

Because relations similar to (6-16), (6-17), and (6-18) still
hold true, we can substitute

$$\frac{m_i'}{m_i} = \frac{V_{(i)}}{V_{(i)} + V_s} \qquad (6\text{-}25)$$

Combining Eqs. (6-24), (6-18), and (6-25), and rearranging, we
obtain

$$m_i = \frac{N_s}{V_{(i)} + V_s} \cdot \frac{1}{\dfrac{A_{is}'\,V_{(i)}}{A_i\,v_{(i)}'} - \dfrac{V_{(i)}}{V_{(i)} + V_s}} \qquad (6\text{-}26)$$

Dividing both the numerator and denominator of the right-hand
side of (6-26) by the ratio $(V_{(i)} + V_s)/V_{(i)}$, we finally obtain

$$m_i = \frac{N_s}{V_{(i)}} \cdot \frac{1}{\dfrac{A_{is}'\,V_{(i)}}{A_i\,v_{(i)}'}\left(1 + \dfrac{V_s}{V_{(i)}}\right) - 1} \qquad (6\text{-}27)$$

In a similar way, an analogous relation for the calculation of
weight fraction can be derived:

$$g_i = \frac{W_s}{W_{(i)}} \cdot \frac{1}{\dfrac{A_{is}'\,W_{(i)}}{A_i\,w_{(i)}'}\left(1 + \dfrac{W_s}{W_{(i)}}\right) - 1} \qquad (6\text{-}28)$$

As with absolute calibration, if weight fractions are to be
calculated, the amounts injected should be weighed. Since in
this case the component of analytical interest is also used as
the standard, relations (6-27) and (6-28) do not require any
correction factor (RMR).

Simple reasoning leads to the realization that the above
version of the standard-addition technique requires, as does
the technique of absolute calibration, at least two chromatographic runs performed under identical conditions with

absolutely defined amounts of sample. In addition, a definite
amount of standard must be added to the initial sample, and
calculation of the sample composition is more complex. However, a detailed analysis shows that the standard-addition
technique offers two significant advantages compared to the
technique of absolute calibration. The former technique allows
calibration chromatograms to be made with material differing
only slightly in composition (mere addition of the pure component to be determined) from the original sample, which may be
important in a number of cases [88]. Preparation of a calibrating solution of the same composition would be difficult
when analyzing complex mixtures. Another advantage is the
possibility of modifying this technique in such a manner that
a knowledge of the absolute amounts injected is unnecessary,
i.e., of using the method of comparison with a reference compound. The step of defined mixing the original sample and
standard, carried out in addition to the procedure adopted for
the absolute-calibration technique, and more complicated calculations are prices which are to be paid for the above-quoted
advantages. If a pure specimen of the component of analytical
interest is not available, it is possible to determine another
component of the mixture analyzed (by the standard-addition
technique) and use the ratio of the corrected peak areas in
the chromatogram of the original mixture for the determination
of the desired component. It must be stated that with the
above variety, the relations for molarity calculations, i.e.,
(6-26) and (6-27), are accurate only if the volumes $V_{(i)}$ and
V_s are additive. This assumption may not always be met to a
sufficient degree. If the deviations from additivity of $V_{(i)}$
and V_s are significant, it is necessary to consider the total
volume of the sample analyzed and standard instead of the sum
of the individual volumes.

2.   *Method Based on Comparison with an Auxiliary Reference Substance Present in the Original Sample*

If the original mixture to be analyzed contains any component which, under operating conditions, is sufficiently separated from the component of analytical interest, then the former can be used as a reference substance if the concentrations of the two components are comparable. The ratios $v_{(i)}/v'_{(i)}$ and $w_{(i)}/w'_{(i)}$ can be then determined from the relation between the peak areas of the component determined and that of the reference substance in the chromatograms of the original and enriched mixtures. This makes it possible to avoid the necessity of knowing the absolute amounts injected. If p denotes the auxiliary reference substance, we may write

$$A_p = C \, RMR_{pr} \, m_p v_{(i)} \tag{6-29}$$

$$A'_p = C \, RMR_{pr} \, m'_p v'_{(i)} \tag{6-30}$$

where the primed symbols refer to a sample enriched by a measured addition of the component of analytical interest (standard). From Eqs. (6-29) and (6-30), it follows that

$$\frac{v_{(i)}}{v'_{(i)}} = \frac{A_p m'_p}{A'_p m_p} \tag{6-31}$$

Since

$$m'_p = \frac{N_p}{V_{(i)} + V_s} \tag{6-32}$$

and

$$m_p = \frac{N_p}{V_{(i)}} \tag{6-33}$$

relation (6-31) can be rewritten as

$$\frac{v_{(i)}}{v'_{(i)}} = \frac{A_p V_{(i)}}{A'_p (V_{(i)} + V_s)} \tag{6-34}$$

Substituting for $v_{(i)}/v'_{(i)}$ from Eq. (6-34) into (6-27) and rearranging, we obtain

$$m_i = \frac{N_s}{V_{(i)}} \frac{1}{(A'_{is} A_p / A_i A'_p) - 1} \qquad (6\text{-}35)$$

The relation for weight fraction can be obtained in an analogous manner:

$$g_i = \frac{W_s}{W_{(i)}} \frac{1}{(A'_{is} A_p / A_i A'_p) - 1} \qquad (6\text{-}36)$$

From the description of the method and from Eqs. (6-35)
and (6-36), it follows that the procedure used is the same as
the method of direct measurement of the charge, except that
the area of the auxiliary reference substance is measured instead of the amounts injected. Neither the kind nor the concentration of the auxiliary reference substance need be known.

3.   Method Based on Comparison with an Added Auxiliary Reference Substance

This method is an analogy to the preceding method, except
that the auxiliary reference substance is not present in the
original sample, but is added to it. The resulting mixture is
then treated in the same way as in the method with an auxiliary
reference substance present in the original sample. Again the
kind and concentration of the reference substance need not be
defined. We need only be concerned for good separation of the
reference substance from the other components of the mixture
in question and for the concentrations of the component determined and of the reference substance to be comparable. It can
be easily seen that also in this case the molarity or weight
fraction of the component determined can be calculated by using

Eqs. (6-35) and (6-36); the quantities $V_{(i)}$ and $W_{(i)}$, however, refer to the original sample under analysis, rather than to the mixture of the sample and the added reference substance. The quantities A'_{is} and A'_p then again refer to the mixture of the original sample and the reference compound, enriched with a defined amount of the standard.

From the concepts of the two techniques using an auxiliary reference compound, it is apparent that the chromatographic runs of the original sample, containing, if need be, an added reference compound, and of the sample enriched by adding a defined amount of the standard need not be carried out under identical conditions. This is another advantage as compared with the method of direct measurement of the amount injected.

4. *Calibration-Curve Methods*

It is not usual to work with a calibration curve when employing the technique of standard addition, in spite of the feasibility of such a mode of work. One of the possible procedures is rooted in plotting known values of m_i or g_i against the respective values of the whole terms on the right-hand sides of the corresponding equations (6-27), (6-28), (6-35), and (6-36). Another procedure is based on the direct use of relations (6-21)-(6-23). Combining these relations and taking into account Eq. (6-25), it is possible to write

$$m_i = \frac{V_{(i)} + V_s}{C\ RMR_{ir}\ V_{(i)} v'_{(i)}}\ A'_{is} - \frac{N_s}{V_{(i)} + V_s} \tag{6-37}$$

Whence, plotting $m_i + [N_s/(V_{(i)} + V_s)]$ against $[(V_{(i)} + V_s)/v'_{(i)}]\ A'_{is}$, we should obtain a straight line which passes through the origin of the coordinate system and has a slope equal to $1/C\ RMR_{ir}$. In the calculation of weight fractions, it is possible to proceed analogously according to the relation

$$g_i = \frac{(W_{(i)} + W_s)M_i}{C \, RMR_{ir} \, W_{(i)}w'_{(i)}} A'_{is} - \frac{W_s}{W_{(i)} + W_s} \tag{6-38}$$

In this case, $g_i + [W_s/(W_{(i)} + W_s)]$ is to be plotted against $[W_{(i)} + W_s)/W_{(i)}w'_{(i)}]A'_{is}$; the slope of the calibration line is given by the expression $M_i/C \, RMR_{ir}$. It is evident that the procedure based on Eqs. (6-37) and (6-38) makes it possible to perform the analysis proper from a single charge $(v'_{(i)}, w'_{(i)})$, i.e., to leave out the injection of the initial sample $(v_{(i)}, w_{(i)})$.

The procedure according to the variant with an auxiliary reference substance can also be utilized for graphical calibration. Substitution for $v'_{(i)}$ in Eq. (6-37) from Eq. (6-30) leads to the following relation:

$$m_i = \frac{RMR_{pr}}{RMR_{ir}} \frac{V_{(i)} + V_s}{V_{(i)}} \frac{A'_{is}}{A'_p} m'_p - \frac{N_s}{V_{(i)} + V_s} \tag{6-39}$$

Analogously, there holds

$$g_i = \frac{RMR_{pr}}{RMR_{ir}} \frac{M_i}{M_p} \frac{W_{(i)} + W_s}{W_{(i)}} \frac{A'_{is}}{A'_p} g'_p - \frac{W_s}{W_{(i)} + W_s} \tag{6-40}$$

Thus, the calibration line can be constructed by plotting $m_i + [N_s/(V_{(i)} + V_s)]$ or $g_i + [W_s/(W_{(i)} + W_s)]$ against $m'_p A'_{is} (V_{(i)} + V_s)/V_{(i)} A'_p$ and $g'_i A'_{is} (W_{(i)} + W_s)/W_{(i)} A'_p$. The respective slopes correspond to the expressions RMR_{pr}/RMR_{ir} and $RMR_{pr}M_i/RMR_{ir}M_p$. The amount of the sample need not be defined when proceeding according to the above variant, but it is necessary either to know the value of m'_p (g'_p), or to keep these values constant both in calibration and in the analysis proper. In the latter case, the values of m'_p or g'_p are involved in the slope of the calibration line

It is apparent from the expressions for the respective slopes that the technique of direct measurement of the sample amount makes it necessary to perform both the calibration and

analysis under identical conditions, whereas the variant with
an auxiliary reference substance provides for carrying out the
two steps under different conditions.

The procedures for obtaining the mathematical expressions
relevant to the individual variants of the standard-addition
technique are not so straightforward as with the other techniques. This is likely why several erroneous relations sug-
tested for calculation when employing this technique can be
found in the literature. This situation has been subjected
to analysis [89].

D.   Internal-Normalization Technique

This technique may be looked upon as a special case of
internal standardization where the role of the internal standard is performed by an arbitrary component of the original
mixture under analysis. The ratio of the amount of the standard and of the sample under analysis can be determined by

$$\frac{N_s}{V_{(i)}} = \frac{d_{(i)} N_s}{M'_{(i)} N_{(i)}} = \frac{d_{(i)}}{M'_{(i)}} \frac{A_s/RMR_{sr}}{\sum_j A_j/RMR_{jr}} \tag{6-41}$$

or

$$\frac{W_s}{W_{(i)}} = \frac{A_s M_s/RMR_{sr}}{\sum_j A_j M_j/RMR_{jr}} \tag{6-42}$$

Combining (6-41) and (6-42) with (6-19) and (6-20), we obtain

$$m_i = \frac{d_{(i)}}{M'_{(i)}} \frac{A_i/RMR_{ir}}{\sum_j A_j/RMR_{jr}} \tag{6-43}$$

$$g_i = \frac{A_i M_i/RMR_{ir}}{\sum_j A_j M_j/RMR_{jr}} \tag{6-44}$$

With regard to (6-5), Eq. (6-43) can be written as

$$x_i = \frac{A_i/RMR_{ir}}{\sum_j A_j/RMR_{jr}} \qquad (6-45)$$

for the calculation of mole fraction. Relations (6-45) and
(6-44) can be arrived at by a simple rearrangement of the definition equations for mole and weight fractions.

It is obvious from the description of the internal-normalization technique that one needs only the chromatogram of the
mixture under analysis, recorded under constant conditions, and
correction factors for the individual components. We need
neither define the absolute amount of sample injected, nor add
any standard. This simplicity is accompanied by several limitations which, in many cases, either completely exclude the
applicability of the internal-normalization technique, or reduce it to a method of approximation of the quantitative composition of sample. Use of the internal-normalization technique is contraindicated in the following cases:

1. A part of the charge is not eluted, i.e., it remains
deposited in the injection-block chamber or in the column, or
is decomposed.

2. A component does not produce any signal for the
detector.

3. Some components of the mixture are not resolved.

4. A component remains unidentified or the respective
correction factor is not known.

5. The area of a peak cannot be measured (peak height
exceeds the recorder scale span).

Because of items 1 and 2, the internal-normalization technique is generally unreliable, since, with a mixture of unknown
composition, one can never be sure that there is not present

some portion not amenable to chromatography or detection. The
possibility of determining any component of a chromatographed
mixture depends upon determining all of the components, and the
precision and accuracy of determination of the individual components are interdependent.

If the correction factors are not known, the internal-normalization technique can serve to estimate the composition of a
given mixture. This type of data correlation raises doubt as
to whether the fractions of uncorrected areas correspond to
mole or weight fractions of the components [8]. From theoretical considerations of Chapter 4, it is obvious that the problem
does not rest with the principle on which the technique has
been based, but is a matter of defining the correction factors.
When using uncorrected areas for the calculation, one cannot
assign any unambiguous concentration units to the area fractions; the results refer to neither mole nor weight fractions.
The extent of deviation from either composition unit depends
upon the detection principle used and on the character of the
compounds detected. Since the RMR value is approximately proportional to molecular weight, the results approach weight data
[11-18]. With the internal-normalization technique, use of
calibration-curve method is out of the question.

E. *Controlled Internal-Normalization Technique*

The degree of uncertainty of results obtained by using
internal normalization can be considerably reduced as follows.
To a sample intended for analysis, a measured amount of some
control substance is added. The substance is chosen such that
the respective correction factor is known and the chromatographic peak of the substance does not coincide with any other
peak in the chromatogram. Let the control substance be denoted
Z, and let Y be that part of the sample under analysis which

which cannot be determined (compare paragraphs 1 to 5 of the preceding technique). The weight fraction of Z, i.e., g_Z in the sample after adding a control substance is given by the relation

$$g_Z = \frac{A_Z f_Z^W}{\sum\limits_{j \neq Y,Z} A_j f_j^W + A_Z f_Z^W + A_Y f_Y^W} \qquad (6\text{-}46)$$

where subscript j denotes any component of the mixture chromatographed except for Z and Y, the correction factors f^W being determined by $f^W = M/RMR$. The weight fraction of component Y in a mixture of the original sample and the control substance, $g_{Y(Z)}$, is given by

$$g_{Y(Z)} = \frac{A_Y f_Y^W}{\sum\limits_{j \neq Y,Z} A_j f_j^W + A_Z f_Z^W + A_Y f_Y^W} \qquad (6\text{-}47)$$

Substituting for $A_Y f_Y^W$ from (6-46) and rearranging,

$$g_{Y(Z)} = 1 - g_Z \frac{\sum\limits_{j \neq Y,Z} A_j f_j^W + A_Z f_Z^W}{A_Z f_Z^W} \qquad (6\text{-}48)$$

The quantity g_Z is

$$g_Z = \frac{W_Z}{W_{(i)} + W_Z} \qquad (6\text{-}49)$$

where W_Z is the weight of the control substance and $W_{(i)}$ is the weight of the original sample analyzed. If g_Y is the weight fraction of the portion Y in the original sample, i.e., before adding Z, then

$$g_Y = \frac{W_Y}{W_{(i)}}$$
(6-50)

and

$$g_{Y(Z)} = \frac{W_Y}{W_{(i)} + W_Z}$$
(6-51)

where W_Y is the weight of Y, and we can write

$$g_Y = \frac{W_{(i)} + W_Z}{W_{(i)}} = \frac{W_Z}{W_{(i)}} \frac{\sum\limits_{j \neq Y,Z} A_j f_j^W + A_Z f_Z^W}{A_Z f_Z^W}$$
(6-52)

When there is any doubt as to whether we are entitled to use
the internal-normalization technique for analysis of a given
sample, one can arrive at a decision by (6-52). The necessary
and sufficient condition is that g_Y be zero. Furthermore, in
this way we may circumvent any limitation of items 1-5 discussed in the section on plain internal normalization.

    If any component of the mixture is separated and detected
during chromatography but remains unidentified, or is identified but its correction factor is unknown, the technique of
controlled internal normalization may serve to determine the
correction factor without the substance being identified and
without having a pure standard. The procedure is as follows.
The chromatographic peak of the substance in question in the
chromatogram of a mixture of the sample analyzed and a control
is taken as the peak of component Y, i.e., the area A_Y is ig-
nored and g_Y is calculated using (6-52). Since

$$g_Y = \frac{A_Y f_Y^W}{\sum\limits_{j \neq Y,Z} A_j f_j^W + A_Y f_Y^W}$$
(6-53)

where A_Y is known, the factor f_Y^W can be expressed by

$$f_Y^W = \frac{\sum\limits_{j \neq Y,Z} A_j f_j^W}{A_Y} \frac{g_Y}{1 - g_Y} \tag{6-54}$$

It now remains to demonstrate the procedure for determining
the other components, i.e., all the components of the original
mixture except for Y. As has been emphasized earlier, when using
the internal-normalization technique the results for the individual components of an analyzed mixture are interdependent. If the
area of the peak of Y, i.e., A_Y, is neglected, all the remaining
data (g_i) will be disfigured. Let these data be denoted g_i^*. It
is obvious that the g_i^*'s would be correct only if they were considered relative to the weight $[W_{(i)} - W_Y]$, i.e., $[W_{(i)} - g_Y W_{(i)}]$,
whereas correct values of g_i refer to the weight $W_{(i)}$. We may
start with a simple balance

$$g_i^* W_{(i)} (1 - g_Y) = g_i W_{(i)} \tag{6-55}$$

from which, after rearranging and substituting for g_Y from
(6-52), it follows that

$$g_i = \frac{A_i f_i^W}{\sum\limits_{j \neq Y,Z} A_j f_j^W} \frac{W_z}{W_{(i)}} \frac{\sum\limits_{j \neq Y,Z} A_j f_j^W}{A_z f_z^W} \tag{6-56}$$

Relation (6-56) obviously corresponds to the internal-standardization technique, where now the control substance Z acquires
the function of an internal standard. However, the terms
$\sum\limits_j A_j f_j^W$ cannot be canceled out if implicit values of the area
fraction $A_i f_i^W / \sum\limits_j A_j f_j^W$ are given. If the mole fraction x is to
be expressed, then

$$x_Y = \frac{N_{(i)} + N_Z}{N_{(i)}} - \frac{N_Z}{N_{(i)}} \frac{\sum\limits_{j \neq Y,Z} A_j f_j^N + A_z f_z^N}{A_z f_z^N} \tag{6-57}$$

$$x_i = \frac{A_i f_i^N}{\sum\limits_{j \neq Y,Z} A_j f_j^N} \frac{N_Z}{N_{(i)}} \frac{\sum\limits_{j \neq Y,Z} A_j f_j^N}{A_Z f_Z} \qquad (6\text{-}58)$$

where $N_{(i)}$ is the total number of moles of the original sample. This procedure is especially advantageous with gases if some portion of the sample can be identical with the carrier gas. The ratios of the numbers of moles can be replaced by volume ratios for gases. The correction factors f^N are given by f^N = 1/RMR.

The technique employing a control substance can give results more reliable than simple internal normalization. However, the former method also suffers from certain limitations. If the problem is the analytical determination of component Y by relation (6-52) or (6-57), this will be possible only in cases where component Y is a significant constituent of the sample analyzed, or, in other words, when the error of determination of the directly determinable portion is considerably smaller than the amount of the component Y. The lesser the content of Y, the greater will be the error involved in its determination. The same also holds true for the determination of the factor f_Y^W. This, of course, arises from the indirectness of the determination. This limitation does not apply to the determination of the other components. The greater the degree of uncertainty in determining component Y, resulting from its low content, the lesser is its influence on the certainty of the other results.

IV. SAMPLE DILUTION

When using highly sensitive detectors, it is sometimes difficult simultaneously to provide for reliable sample injection and ensure that the concentration of the substance chromatographed

in the column effluent will not exceed the limit of detector-
response linearity. It is of advantage, in such cases, to
appropriately dilute the sample before injecting it into the
chromatograph. The sample may be diluted in several steps, so
that the volume $V_{(i)0}$ of the initial sample is diluted with a
solvent to $V'_{(i)1}$; from the latter, a volume $V_{(i)1}$ is taken
which, again, is diluted to $V'_{(i)2}$, etc. If m_{i0} is the molarity
of i in the initial sample, then the molarity after the nth
dilution, m_{in}, is given by

$$m_{in} = m_{i0} \frac{V_{(i)0} V_{(i)1} \cdots V_{(i)(n-1)}}{V'_{(i)1} V'_{(i)2} \cdots V'_{(i)n}} = m_{i0} F_{mi} \qquad (6\text{-}59)$$

Similarly, for the weight fraction

$$g_{in} = g_{i0} \frac{W_{(i)0} W_{(i)1} \cdots W_{(i)(n-1)}}{W'_{(i)1} W'_{(i)2} \cdots W'_{(i)n}} = g_{i0} F_{gi} \qquad (6\text{-}60)$$

The terms F_{mi} and F_{gi} will be called dilution factors. In an
analogous manner, for a standard, or better said, for a solution of the standard after the ℓth dilution,

$$m_{s\ell} = m_{s0} F_{ms} \qquad (6\text{-}61)$$

$$g_{s\ell} = g_{s0} F_{gs} \qquad (6\text{-}62)$$

The volume and weight of the total and of the injected amount
of the standard solution will be denoted by the same symbols
as used previously, i.e., $V_{(s)}$, $W_{(s)}$, $v_{(s)}$, and $w_{(s)}$. For the
total amount of standard in a given amount of standard solution
after the ℓth dilution

$$N_{s\ell} = m_{s\ell} V_{(s)\ell} = m_{s0} F_{ms} V_{(s)\ell} \qquad (6\text{-}63)$$

$$W_{s\ell} = g_{s\ell} W_{(s)\ell} = g_{s0} F_{gs} W_{(s)\ell} \qquad (6\text{-}64)$$

The introduction of the dilution factor is, in fact, a

generalization of the techniques discussed in the preceding section, which, of course, is being reflected in the respective mathematical relations. This generalization will be dealt with in the following section.

A. Absolute-Calibration Technique

1. Direct-Comparison Method

Since

$$m_{in} = \frac{RMR_{sr}\, v_{(s)\ell}\, A_i}{RMR_{ir}\, v_{(i)n}\, A_s}\, m_{s\ell} \tag{6-65}$$

then, with respect to (6-59) and. (6-61), we may write

$$m_{i0} = \frac{RMR_{sr}\, v_{(s)\ell}\, A_i\, F_{ms}}{RMR_{ir}\, v_{(i)n}\, A_s\, F_{mi}}\, m_{s0} \tag{6-66}$$

In an analogous manner, we obtain for the weight fraction

$$g_{i0} = \frac{RMR_{sr}\, M_i\, w_{(s)\ell}\, A_i\, F_{ms}}{RMR_{ir}\, M_s\, w_{(i)n}\, A_s\, F_{mi}}\, g_{s0} \tag{6-67}$$

2. Calibration-Curve Method

Consider the most generalized procedures involving corrected peak areas (Fig. 8). From the calibration graphs prepared in this manner, and for the values A_i/RMR_{ir} and $A_i M_i/RMR_{ir}$ found, we can directly read $v_{(i)n} m_{i0} F_{mi}$ or $w_{(i)n} g_{i0} F_{gi}$, from which the concentrations m_{i0} and g_{i0} can be easily determined.

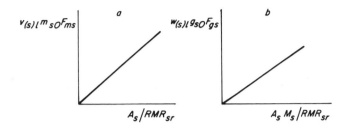

FIG. 8. Calibration graphs for the determination of molarity (a) and weight fraction (b) by absolute calibration with sample dilution [$v_{(s)l}$, $w_{(s)l}$, volume and weight of injected calibration solution after lth dilution; m_{s0}, g_{s0}, molarity and weight fraction of standard in initial calibration solution; F_{ms}, F_{gs}, dilution factors given by Eqs. (6-55) and (6-56); for other symbols see Fig. 6].

B.   Internal-Standardization Technique

1.   Direct-Comparison Method

Here we shall distinguish between two alternatives:

1. The standard is added to the sample and the mixture is diluted. The dilute mixture is injected into the chromatograph. Since the absolute amount of sample injected has no bearing in the internal-standardization technique, the procedure and the interpretation of the chromatogram are the same as those without a previous dilution step.

2. The sample to be analyzed and the standard solution are separately diluted, and the resulting solutions are mixed. The mixture is injected into the chromatograph. In this case, it is of advantage to use (6-19) directly, by which

$$m_{in} = \frac{N_{sl}\,RMR_{sr}\,A'_i}{V_{(i)n}\,RMR_{ir}\,A'_s} \tag{6-68}$$

Substituting for m_{in} and N_{sl} from (6-59) and (6-63) and rearranging,

$$m_{i0} = \frac{m_{s0}\,V_{(s)l}\,RMR_{sr}\,A'_i F_{ms}}{V_{(i)n}\,RMR_{ir}\,A'_s F_{mi}} \tag{6-69}$$

and, in an analogous way,

$$g_{i0} = \frac{g_{s0}\,W_{(s)l}\,RMR_{sr}\,M_i A'_i F_{gs}}{W_{(i)n}\,RMR_{ir}\,M_s A'_s F_{gi}} \tag{6-70}$$

2. *Calibration-Curve Method*

As a parallel to alternative 1, there are features which are identical to the direct-comparison method. For the most general case, corresponding to 2, the graphs in Fig. 9 apply.

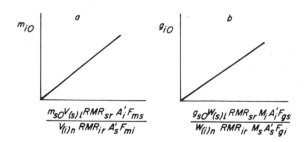

FIG. 9. Calibration graphs for the determination of molarity (a) and weight fraction (b) by the technique of internal standardization [$V_{(s)l}$, $W_{(s)l}$, volume and weight of standard solution after lth dilution, added to sample volume $V_{(i)n}$ or weight $W_{(i)n}$ after nth dilution; M_i, M_s, molecular weight of component and standard; for other symbols see Fig. 7].

C.   Standard-Addition Technique

1.   Method of Direct Sample Measurement

In practice, those procedures are of a greater significance where the sample to be analyzed and the standard are diluted independently of each other. The respective solutions are then treated in the same way as in procedures without previous dilution. With regard to (6-27), we may write

$$m_{in} = \frac{N_{s\ell}}{V_{(i)n}} \frac{1}{\dfrac{A'_{is}v_{(i)n}}{A_i v'_{(i)n}} \left(1 + \dfrac{V_{(s)\ell}}{V_{(i)n}} \right) - 1} \qquad (6\text{-}71)$$

which, upon substituting for m_{in} and $N_{s\ell}$ from (6-59) and (6-63) yields

$$m_{i0} = \frac{m_{s0}V_{(s)\ell}F_{ms}}{W_{(i)n}F_{mi}} \frac{1}{\dfrac{A'_{is}v_{(i)n}}{A_i v'_{(i)n}} \left(1 + \dfrac{V_{(s)\ell}}{V_{(i)n}} \right) - 1} \qquad (6\text{-}72)$$

For the weight fraction,

$$g_{i0} = \frac{g_{s0}W_{(s)\ell}F_{gs}}{W_{(i)n}F_{gi}} \frac{1}{\dfrac{A'_{is}w_{(i)n}}{A_i w'_{(i)n}} \left(1 + \dfrac{W_{(s)\ell}}{W_{(i)n}} \right) - 1} \qquad (6\text{-}73)$$

2.   Method of Comparison with an Auxiliary Reference Substance Present in the Initial Sample

For the same dilution procedure as that used in the method of direct measurement of the sample injected, we can write immediately

$$m_{i0} = \frac{m_{s0}V_{(s)\ell}F_{ms}}{V_{(i)n}F_{ni}} \frac{1}{(A'_{is}A_p / A_i A'_p) - 1} \qquad (6\text{-}74)$$

and

$$g_{i0} = \frac{g_{s0} W_{(s)} \ell^F gs}{W_{(i)n} F_{gi}} \frac{1}{(A'_{is} A_p / A_i A'_p) - 1}$$
(6-75)

3. *Method of Comparison with an Added Auxiliary Reference Substance*

The relationship between this method and the method of comparison with an auxiliary reference substance present in the original sample is analogous to the corresponding relationship between the methods of analysis without any dilution. It follows that, in this case, relations (6-74) and (6-75) hold true. However, if $V_{(i)n}$ and $W_{(i)n}$ are the actual volume and weight of the final sample/reference-compound/diluent mixture, which is to be further mixed with the standard solution ($V_{(s)\ell}$ or $W_{(s)\ell}$), an additional factor accounting for the dilution of the original sample with the reference compound must be applied along with F_{ni} or F_{gi}. This factor is given by the ratio of the volume or weight of the sample/reference-compound solution (before adding the diluent) to that of the sample alone.

With the standard-addition technique involving sample dilution, the potentialities of working with a calibration curve are the same as with the original versions, i.e., without the diluting procedure. In order to obtain the respective relations, the products $m_{i0} F_{mi}$, $g_{i0} F_{gi}$, and $m_{s0} F_{ms}$, $g_{s0} F_{gs}$ are introduced instead of the quantities m_i, g_i, and m_s, g_s, respectively, and the subscripts (i) and (s) accompanying the quantities V, v, W, and w are replaced by the subscripts (i)n and (s)ℓ.

D. *Internal-Normalization Technique*

In this technique, no data on absolute amounts of the treated material are used for calculations. Dilution, therefore, influences neither the procedure nor the calculations.

Thus, relations (6-44) and (6-45) are relevant as derived above. An obvious requirement is that the peak of solvent must not co-incide with any other peak in the chromatogram, and shall not be taken into account in evaluating the chromatogram.

E. *Controlled Internal-Normalization Technique*

If dilution is performed after adding the control substance, the procedure remains the same as in cases without dilution. The solvent peak is not considered. If, on the other hand, the sample to be analyzed and the control substance have been diluted independently before being blended, then

$$g_{io} = \frac{A_i f_i^W}{\sum\limits_{j \neq Y, Z, d} A_j f_j^W} \frac{W_{(Z)\ell}}{W_{(i)n}} \frac{\sum\limits_{j \neq Y, Z, d} A_j f_j^W}{A_Z f_Z^W} \frac{F_{gZ}}{F_{gi}} \tag{6-76}$$

where d denotes the solvent. The calculation of mole fraction would again be practicable only in the analyses of gases. The relation is

$$x_{io} = \frac{A_i f_i^N}{\sum\limits_{j \neq Y, Z, d} A_j f_j^N} \frac{N_{(Z)\ell}}{N_{(i)n}} \frac{\sum\limits_{j \neq Y, Z, d} A_j f_j}{A_Z f_Z^N} \frac{F_{mZ}}{F_{mi}} \tag{6-77}$$

For the gases the ratio of mole numbers, $N_{(Z)\ell}/N_{(i)n}$, can be replaced by that of their volumes, i.e., $V_{(Z)\ell}/V_{(i)n}$.

If both the sample and standard are much diluted with the same solvent, then both solutions have practically the same density, and their weight ratio is given by the volume ratio. This fact can be utilized to advantage whenever it is necessary to make calculations with the ratio of the weights of standard and sample in question, regardless of whether the weights refer to the bulk sample or to the amount injected.

V. EXPERIMENTAL DETERMINATION OF CORRECTION FACTORS

Relations (4-64) and (6-10) are the basis for any method of experimental determination of the correction factors, though the actual procedures can differ according to the working technique employed. Several nonchromatographic techniques for the determination of the response factors have been developed; their use, however, requires special calibrating devices. Besides these techniques, the techniques described above, i.e., absolute calibration, internal standardization, and internal normalization, also can be used to advantage. The respective procedures involve the interpretation of the chromatogram of a mixture of known composition (both qualitative and quantitative) and follow directly from the respective mathematical formulations. The techniques of absolute calibration and internal standardization yield the ratio RMR_{ir}/RMR_{sr}. Since the relations

$$RMR_{ir} = \frac{MR_i}{MR_r} \qquad (6\text{-}78)$$

and

$$RMR_{sr} = \frac{MR_s}{MR_r} \qquad (6\text{-}79)$$

hold true, where MR stands for the molar response, we can write

$$\frac{RMR_{ir}}{RMR_{sr}} = \frac{MR_i}{MR_s} = RMR_{is} \qquad (6\text{-}80)$$

Thus, the ratio RMR_{ir}/RMR_{sr} corresponds to the relative molar response for substance i, with the standard playing the role of the reference substance. If the relative molar response for substance i is to be expressed relative to some other substance, i.e., generally, relative to substance r, it is necessary to treat the compound r in the same way as compound i, whereby we obtain RMR_{rs}. The value of RMR_{ir} is then determined from

$$\frac{MR_i}{MR_s} \frac{MR_s}{MR_r} = \frac{RMR_{is}}{RMR_{rs}} = RMR_{ir} \qquad (6\text{-}81)$$

If the standard is identical to the compound r, then RMR_{ir}/RMR_{sr} gives directly the value of RMR_{ir}, since RMR_{sr} (RMR_{rr}) equals unity.

When using the internal-normalization technique, it is possible to proceed in several ways. In general, it is neces-sary to prepare n + 1 mixtures of different quantitative compo-sitions if there are n components of unknown concentrations present in the mixtures, provided the concentration of at least one component is always known, no matter of which one. Whence, if a mixture of known concentrations for all the components is available, it is sufficient to run only one chromatogram. If, on the other hand, only one of the components in question is available as a pure standard, it is possible to base the calcu-lation on the chromatograms of n mixtures of different contents of the above component, where n is the total number of the com-ponents of the mixtures. In any case, however, it is necessary to solve a set of as many equations as there are components in the mixture. The technique of controlled internal normaliza-tion, performed through use of the computational procedure we have outlined, theoretically provides for the determination of response factors without knowing the content of any component of the original mixture. Of course, in any mode of indirect determination, the primary calculation can be carried only on a weight basis, except for the work with gases. The above concepts are the basis of Janik's linear relationship method [90-93].

Chapter 7

SPECIAL TECHNIQUES

There are cases in which it is necessary to separate the components of analytical interest from an excess ballast material prior to the chromatographic analysis proper and/or to accumulate traces of these components up to amounts easily detectable and determinable by conventional gas-chromatographic means. These procedures are sometimes combined with chemical derivations of the components under determination. The procedure proper to be chosen for solving a particular analytical problem depends on the physical state of the sample and the nature and concentration of the components. The sample amount available is also an important factor in this respect. When applying the methods of this type, the most suitable units for expressing the analytical results are those of weight per unit volume. Hence, these units will be employed throughout this chapter.

I. TRAPPING OF COMPONENTS OF GASEOUS SAMPLES

*A. Accumulation of the Components Without Altering
     Their Original Proportions*

A common and straightforward method of the isolation and
accumulation of components of gaseous mixtures consists in
drawing a defined volume of the latter through a tube packed
with a suitable sorbent [94-97]. During this procedure, the
components being trapped are subject to frontal chromatography;
the fronts of the individual components migrate down the tube
at a velocity determined by the sorption capacity of the packing and the velocity of the gas. Hence, if the proportions of
the components entrapped in the sampling tube are to be the
same as in the analyzed gas, it is necessary to stop the suck-
ing before the front of the least sorbed component breaks
through the tube. The degree of accumulation can, within wide
limits, be controlled by the choice of the sorbent and by the
temperature of the sampling tube.

An interesting alternative to this method is the accumulation of substances in the so-called temperature-gradient tube,
introduced by Kaiser [98]. In this case, a steep temperature
gradient across the tube is maintained during the sampling,
the flow of the gas being oriented in the direction of temperature drop. This results in a partial separation of the components being accumulated, which suppresses the possibility of
chemical reactions occurring between the components.

The problem of transferring the concentrate from the sampling tube into the gas chromatograph is practically the same
with both versions of this kind of sampling described herein.
Basically, the transfer consists in inserting the tube into
the path of the carrier gas ahead of the column, warming the
tube up to an appropriate temperature, and then purging the
desorbed substances by the carrier gas onto the column.

For these purposes, it is appropriate to construct the sampling
tube as a probe which can be provided with an injection needle
on one end, install an oven above the sample-inlet port of the
gas chromatograph, and insert a three-way stopcock into the
carrier-gas line ahead of the inlet port. The stopcock, being
connected through its two ports into the line, provides for directing the carrier-gas stream either normally into the GC column or into the probe, connected by a metal capillary to the
third port of the stopcock.

The procedure proper of purging the concentrate into the
gas chromatograph is as follows: The initial setting of the
stopcock is such that the carrier gas is allowed to flow directly into the column, and the oven is warmed up to the desired
temperature. Under this arrangement, the sampling probe is
connected to the free port of the stopcock and inserted into
the oven so as to keep the packing within the heated zone, and
the injection needle of the probe is stuck into the inlet-port
septum just far enough to push the needle tip through about
half the thickness of the septum. With this setting, the entire inner space of the probe is virtually closed. As soon as
the sorbent has been warmed up, which usually takes some min-
utes, the stopcock is turned into the position directing the
carrier-gas stream into the probe, and the needle of the latter
is pierced through the septum by pushing the probe down. After
the desorbed components have been purged into the gas chromatograph and the chromatogram run, the stopcock is turned back
into the initial position and the probe is withdrawn and
disconnected.

With the technique of the temperature-gradient tube, the
desorption is carried out under a temperature gradient, by the
carrier gas streaming in the direction of decreasing temperature.

The techniques employing a concentration tube are currently
used also in combination with capillary gas chromatography

[99-104]. However, owing to the extremely low inner volume of
the capillary column, special provisions have to be made in this
case in order to avoid intolerable loss of the separation efficiency upon the transfer of the desorbate from the tube into the
column. The following procedure has been well tried: An inlet
part of the capillary column (or the whole column) is cooled in
solid carbon dioxide or liquid nitrogen during purging of the
concentrate from the sampling tube into the column, so that the
components to be determined are frozen just at the inlet while
the excess gas passes by. After this step is completed, the
system is reset to normal operation, and the components introduced are chromatographed, usually at programmed temperature.
Arrangements for performing this kind of work are available com-
mercially as options to standard GC equipment.

The quantitation of the chromatogram of the concentrate is
carried out by the absolute-calibration method, i.e., by chromatographing separately a defined amount of a standard substance
and comparing the peak areas in both chromatograms. The calculation of the results can be carried out as follows: Let the
volume $V_{(i)}$ of the gas drawn through the tube contain the weight
amount W_i of component i, i.e., $q_i(T, P) = W_i/V_{(i)}$, where the
symbols T and P are to remind that q_i is a function of the temperature and pressure of the gas. The amount W_i is entrapped
in the tube and eventually introduced into the gas chromatograph,
so that

$$W_i = CA_i f_i^W \tag{7-1}$$

In the calibration run, we inject a volume $v_{(s)}$ of the standard
sample in which the concentration of the standard compound is
q_s; hence,

$$w_s = q_s v_{(s)} = CA_s f_s^W \tag{7-2}$$

Dividing Eq. (7-1) by Eq. (7-2), solving for W_i, and replacing
the latter by $q_i V_{(i)}$ gives

$$q_i(T, P) = \frac{A_i f_i^W}{A_s f_s^W} \frac{q_s v_{(s)}}{V_{(i)}} \tag{7-3}$$

In work with gases, it may be more convenient to express
the concentrations of the components as mole fractions, the
values of which are independent of T and P. The mole fraction
of component i in the gas analyzed is $x_i = N_i/N_{(i)}$, where N_i
and $N_{(i)}$ are the number of moles of component i and the total
number of moles of the gaseous sample, respectively. The
quantities N_i and $N_{(i)}$ can be expressed as W_i/M_i and $PV_{(i)}/RT$,
so that

$$x_i = \frac{q_i RT}{M_i P} \tag{7-4}$$

and the combination of Eqs. (7-4) and (7-3) gives

$$x_i = \frac{A_i f_i^W}{A_s f_s^W} \frac{q_s v_{(s)}}{V_{(i)}} \frac{RT}{PM_i} \tag{7-5}$$

One-hundred times x_i is the volume percentage, and 10^6 times x_i
is the number of parts per million of component i in the gas
analyzed.

In addition to the procedure just described, liquid extraction of the sampling-tube packing was also recommended in order
to recover the concentrate [105].

B.   Chromatographic-Equilibration Method [106-108]

The technique involving the use of the sampling tube can
be used advantageously while utilizing the principles of frontal

chromatography. In contrast to the procedure described in the
preceding section, i.e., stopping the suction of the sample
through the tube before the front of the least sorbed component
reaches the outlet of the latter while measuring the sample
volume, the sample is drawn through the tube until the front of
the most sorbed component has broken through. Under these circumstances, the concentrations of the components in the sorbent
are equilibrated with the corresponding concentrations in the
gas being drawn over the sorbent; if the temperature of the
tube and the composition of sampled gas are constant, the composition of the gas leaving the tube is the same as that of the
gas entering it, and the concentrations of the components
sorbed in the tube remain unchanged upon further drawing the
gas through the tube. This implies that the volume of the gas
sampled is not an important quantity and need not be measured.
On the other hand, it is necessary to know the partition coefficients of the individual components under the conditions in
the sorption system and the amount of the sorbent in the tube.

     In contradistinction to the previously described technique,
sampling by the method of chromatographic equilibration dis-
criminates between the individual components as to the extent
of the degree of their accumulation: the larger the partition
coefficient of the substance entrapped, the higher the degree
of its accumulation. This feature is advantageous; very often
there is a proportionality between the concentrations and the
volatilities of the components in the gas analyzed, but, in a
properly chosen sorption system, the respective partition coefficients would decrease with increasing volatilities of the
components, so that some equalization of the amounts of the
individual components occurs automatically during the process
of their accumulation. This discriminating effect can further
be controlled by the selectivity of the sorbent.

     If the concentrations of the components in the analyzed

sample are sufficiently low, then their equilibrium concentrations in the sorbent will also be low, and the sorbed components will neither appreciably interact with each other nor
alter the sorption properties of the sorbent proper. Hence,
the partition coefficients controlling the degree of accumulation in this multicomponent system of the mixture of solutes
and the sorbent will not differ appreciably from the partition
coefficients measured in the corresponding binary solute-sor-
bent systems under the same conditions. These partition coefficients can easily be determined from the retention data
measured by conventional elution chromatography on a column
packed with the same sorbent material as that used in the
sampling tube.

1. *Calculation of Results*

In the packing of the sampling tube, the sorbate components are present both in the sorbent phase and in the void
spaces between the sorbent particles; also, some amount of the
components is present in the unpacked parts of the probe. However, the sorption equilibrium is represented merely by the
content present in the sorbent proper, so that the portion
occurring in the void volumes, which obviously cannot be eliminated, presents a certain complication of the situation.

Let the overall weight amount of component i present in
the sorbent and in the gaseous phases within the probe be W_{iT};
this amount is purged into the gas chromatograph and determined by absolute calibration as

$$W_{iT} = \frac{A_i f_i^W}{A_s f_s^W} q_s v_{(s)} \tag{7-6}$$

The partition coefficient of component i in the sorbent packing
of the tube, K_{iSG}, is

$$K_{iSG} = \frac{q_{iS}}{q_{iG}} \tag{7-7}$$

where q_{iS} and q_{iG} are the equilibrium concentrations of component i in the sorbent and in the gaseous phase; these concentrations are given by

$$q_{iS} = \frac{W_{iS}}{V_S} \tag{7-8}$$

$$q_{iG} = \frac{W_{iGT}}{V_{GT}} \tag{7-9}$$

where W_{iS} and W_{iGT} are the weight amounts of component i contained in the volumes V_S and V_{GT} of the sorbent and the gaseous phase in the sampling tube, respectively. In our case, the quantity to be determined is the concentration q_{iG}, which, with regard to Eqs. (7-7) and (7-8), can be rewritten as

$$q_{iG} = \frac{W_{iS}}{V_S K_{iSG}} \tag{7-10}$$

At equilibrium,

$$W_{iS} = W_{iT} - W_{iGT} \tag{7-11}$$

where [cf. Eqs. (7-7), (7-8), and (7-9)]

$$W_{iGT} = \frac{W_{iS} V_{GT}}{K_{iSG} V_S} \tag{7-12}$$

Combining Eqs. (7-11) and (7-12) gives

$$W_{iS} = \frac{W_{iT}}{1 + (V_{GT}/K_{iSG} V_S)} \tag{7-13}$$

which, further combined with Eq. (7-10), results in

$$q_{\underline{i}G} = \frac{W_{i\underline{T}}}{K_{\underline{i}SG}V_S + V_{GT}} \tag{7-14}$$

Let us recall that V_S and V_{GT} denote the volumes of the sorbent proper in the sampling tube and of the entire void space inside the probe, respectively, $W_{i\underline{T}}$ being determined by Eq. (7-6). In most cases, $K_{\underline{i}SG}V_S \gg V_{GT}$.

Equation (7-14) can be simplified by expressing $K_{\underline{i}SG}$ in terms of chromatographic retention data. Let us consider the sampling tube as a column in which component i is subject to frontal chromatography; we could as well imagine the elution chromatography of component i to proceed in the tube. In both cases, the basic quantity characterizing the chromatographic retention is the retention time. In frontal chromatography, it is the time that is necessary under the given conditions for the concentration inflection of the front to pass through the tube, while in elution chromatography this corresponds to the migration time of the concentration maximum of the elution zone. Hence, under the circumstances discussed above, the situation which is established by virtue of frontal chromatography can be characterized in terms of retention data measured by elution chromatography in the corresponding system. For our purposes, it is more convenient to calculate with the retention volume, V_{Ri}, rather than with the retention time, the former being given by the product of the retention time and the pressure-averaged volume flow rate of the gas drawn through the tube. In our case, the retention volume is

$$V_{Ri} = V_{GT} + K_{\underline{i}SG}V_S \tag{7-15}$$

which, combined with Eq. (7-14) gives

$$q_{\underline{i}G} = \frac{W_{\underline{i}\underline{T}}}{V_{Ri}} \tag{7-16}$$

The retention volume is related to the specific retention volume, V_{gi}, by [109]

$$V_{gi} = \frac{V_{Ri} - V_{\underline{GT}}}{W_{\underline{S}}} \frac{273}{T} \tag{7-17}$$

where $W_{\underline{S}}$ is the weight of the sorbent in the tube and T is the absolute temperature at which the sorbent is kept during sampling. Equations (7-17) and (7-16) can be combined to give

$$q_{\underline{i}G} = \frac{W_{\underline{i}\underline{T}}}{V_{gi} W_{\underline{S}} (T/273) + V_{\underline{GT}}} \tag{7-18}$$

The value of V_{gi} can of course be determined by measurement on a column containing any suitable amount of the sorbent; after replacing the quantities V_{Ri}, $V_{\underline{GT}}$, and $W_{\underline{S}}$ in Eq. (7-17) by V^*_{Ri}, $V_{\underline{m}}$, and $W^*_{\underline{S}}$, where V^*_{Ri} is the retention volume as measured on the weight amount $W^*_{\underline{S}}$ (generally different from $W_{\underline{S}}$) of the sorbent in the column and $V_{\underline{m}}$ is the dead retention volume, the equation is applicable quite generally.

The advantage of calculating with V_{gi} rests with the fact that log V_{gi} is linearly proportional to the inverse of absolute temperature of the sorbent within certain limits, i.e.,

$$\log V_{gi} = \frac{A}{T} - B \tag{7-19}$$

where A and B are constants. Thus, the specific retention volume at the temperature of sampling can be determined by extrapolation or interpolation by Eq. (7-19) from V_g values measured at different temperatures.

II.   EXTRACTION OF LIQUID SAMPLES BY LIQUID EXTRAHENTS

Upon a single extraction, the amount W_i of component i originally present in the sample is divided between the parent solution and the extract (subscripts p and e, respectively) according to the equation

$$W_i = W_{ip} + W_{ie} \qquad (7-20)$$

If the volumes of the parent solution and of the extract are $V_{(i)p}$ and $V_{(i)e}$, Eq. (7-20) can be rewritten as

$$W_i = q_{ip} V_{(i)p} + q_{ie} V_{(i)e} \qquad (7-21)$$

Provided that the amounts of the extractable components and the mutual miscibility of the sample and the extrahent are negligible, the volumes $V_{(i)p}$ and $V_{(i)e}$ are practically identical to the volumes of the sample and the extrahent, $V_{(i)}$ and V_e, respectively. The distribution of component i between both phases is determined by the partition coefficient, which can be defined as $K_{ipe} = q_{ip}/q_{ie}$. Hence, Eq. (7-21) can further be rewritten to read

$$W_i = q_{ie} [K_{ipe} V_{(i)p} + V_{(i)e}] \qquad (7-22)$$

or

$$W_i = W_{ie} \left[\frac{K_{ipe} V_{(i)p}}{V_{(i)e}} + 1 \right] \qquad (7-23)$$

The concentration of component i in the initial sample, $q_i = W_i/V_{(i)}$, is then given by

$$q_i = \frac{W_{ie}}{V_{(i)}} \left[\frac{K_{ipe} V_{(i)p}}{V_{(i)e}} + 1 \right] \qquad (7-24)$$

A. *Direct Analysis of the Extract*

The concentrations of the extractants in the extract can be determined by employing conventional techniques of quantitative gas chromatography as described in Chapter 6. However, if these concentrations are to be recalculated to the corresponding concentrations in the original sample, it is necessary to know the factor $\{[K_{ipe}V_{(i)p}/V_{(i)e}] + 1\}$. The value of this factor is obviously a function of the composition and temperature of the sample-extrahent system. Let us call this quantity the "system factor" and denote it f_{ipe}. Thus, if just the extract is subject to GC analysis, the combinations of Eq. (7-24) with the definitions of the individual analytical techniques result in the following relations.

1. *Absolute-Calibration Technique*

$$q_i = \frac{A_{ie}}{A_s}\ \frac{f_i^W}{f_s^W}\ \frac{v_{(s)}}{v_{(i)e}}\ \frac{V_{(i)e}}{V_{(i)}}\ f_{ipe}q_s \tag{7-25}$$

2. *Internal-Standard Technique*

In this technique, as well as in that of standard addition, either the entire volume of the extract, $V_{(i)e}$, or a part of it may be employed in the procedure of quantitation proper. Thus, denoting the volume of the extract mixed with the standard $V_{(i)e}^o$, we have

$$q_i = \frac{W_s}{V_{(i)e}^o}\ \frac{V_{(i)e}}{V_{(i)}}\ \frac{A'_{ie}}{A'_s}\ \frac{f_i^W}{f_s^W}\ f_{ipe} \tag{7-26}$$

Equation (7-26) involves the addition of the standard to a given amount of the extract separated from the system. However, it is also possible to add the standard substance directly to the original sample prior to its extraction. If need be, the standard may be added to the extrahent or to the sample-extract system, which, of course, renders the same situation as if the standard were added to the sample. Provided the extraction is carried out after the standard compound has been added to the sample, we can write for the total weights of the component to be determined (W_i) and of the standard (W_s) in the sample-extract system [cf. Eq. (7-23)]:

$$W_i = W_{ie}\left[\frac{K_{ipe}V_{(i)p}}{V_{(i)e}} + 1\right] \qquad (7\text{-}27)$$

and

$$W_s = W_{se}\left[\frac{K_{spe}V_{(i)p}}{V_{(i)e}} + 1\right] \qquad (7\text{-}28)$$

where W_{ie} and W_{se} are the total weights of component i and of the standard present in the extract of volume $V_{(i)e}$, $V_{(i)p}$ being the volume of the parent material. Dividing Eq. (7-27) by Eq. (7-28) and denoting the expressions in the brackets by f_{ipe} and f_{spe}, respectively, we obtain

$$\frac{W_i}{W_s} = \frac{W_{ie}f_{ipe}}{W_{se}f_{spe}} . \qquad (7\text{-}29)$$

Running the chromatogram of a sample of the extract will give peak areas A'_{ie} and A'_{se}, for which

$$\frac{W_{ie}}{W_{se}} = \frac{A'_{ie}f_i^W}{A'_{se}f_s^W} \qquad (7\text{-}30)$$

so that combining Eqs. (7-29) and (7-30) and replacing W_i by $q_i V_{(i)}$ gives

$$q_i = \frac{W_s}{V_{(i)}} \frac{A'_{ie}}{A'_{se}} \frac{f_i^W}{f_s^W} \frac{f_{i\underline{pe}}}{f_{s\underline{pe}}} \tag{7-31}$$

With a judicious choice of the standard compound the system factors can be cancelled.

3. *Standard-Addition Technique: Direct Measurement of the Sample Charge*

$$q_i = \frac{W_s}{V_{(i)\underline{e}}^o} \frac{V_{(i)\underline{e}}}{V_{(i)}} \frac{f_{i\underline{pe}}}{\dfrac{A'_{is\underline{e}}}{A_{i\underline{e}}} \dfrac{V_{(i)\underline{e}}}{V'_{(i)\underline{e}}} \left(1 + \dfrac{V_s}{V_{(i)\underline{e}}^o}\right) - 1} \tag{7-32}$$

4. *Standard-Addition Technique: Comparison with an Auxiliary Reference Substance*

$$q_i = \frac{W_s}{V_{(i)\underline{e}}^o} \frac{V_{(i)\underline{e}}}{V_{(i)}} \frac{f_{i\underline{pe}}}{(A'_{is\underline{e}}/A_{i\underline{e}})(A_{p\underline{e}}/A'_{p\underline{e}}) - 1} \tag{7-33}$$

5. *Internal-Normalization Technique*

$$\frac{q_i}{\sum\limits_{j} q_j} = \frac{W_{i\underline{e}} f_{i\underline{pe}}}{\sum\limits_{j} W_{j\underline{e}} f_{j\underline{pe}}} = \frac{A_{i\underline{e}} f_i^W f_{i\underline{pe}}}{\sum\limits_{j} A_{j\underline{e}} f_j^W f_{j\underline{pe}}} = g_i \tag{7-34}$$

and q_i is given by $q_i = g_i d_{(i)}$

A common disadvantage of all the alternatives described by Eqs. (7-25), (7-26), and (7-31)-(7-34) is the necessity of knowing the system factor. In special cases, viz., when components chemically very dissimilar to the parent material are extracted with a solvent similar to the components, the value of K_{ipe} may be small enough to render the term $K_{ipe} V_{(i)p} / V_{(i)e}$ neglibible in comparison with unity, so that the value of f_{ipe} would approach unity. A typical example of this is the extraction of hydrocarbons from aqueous solutions with a hydrocarbon extrahent [110].

B. *Techniques Involving the Elimination of the System Factor*

The extraction step apparently brings a special problem into GC quantitation: a problem the principles of which lie in the field of the theory of solutions. It is not intended to discuss here the problems of extraction equilibria or the selection of suitable extracting agents; the aim is rather to show the analytical implications of these problems and how they are to be taken into account in formulating the practical procedures of quantitative GC analysis.

1. *Use of a Reference Model Mixture*

Theoretically, the procedures as described by Eqs. (7-25), (7-26), and (7-31)-(7-33) can be employed in such a way that the system factor would be definitely eliminated. Namely, when performing the complete procedure with a model mixture having a known content of the substance under determination and comparing the measured data with those obtained by processing under identical conditions the sample analyzed, the concentration q_i can be determined without knowing the system factor, as the latter is cancelled in the calculation of results. However, this will be so only if the overall compositions of the systems with the sample under analysis and with the model mixture (inclusive of the extrahent in both cases) are the same, and if both systems are kept at the same temperature (the effect of pressure on the partition coefficient in a liquid-liquid

system is negligible). Actually, the above-outlined processing of the model mixture gives all the data necessary to determine the value of the system factor and even of the partition
coefficient proper. Consideration of the partition coefficient
instead of the system factor in the calculation of q_i affords
some generalizations, viz., when calculating directly with
$K_{\underline{i}\underline{p}\underline{e}}$, it would theoretically be unnecessary to keep the ratio
$V_{(i)\underline{p}}/V_{(i)\underline{e}}$ invariant, but, practically, this condition would
hold only provided that $K_{\underline{i}\underline{p}\underline{e}}$ is independent of the $V_{(i)\underline{p}}/V_{(i)\underline{e}}$
ratio within the range of variation of the ratio. As this can
not generally be guaranteed, it is better from the analytical
viewpoint to adhere to the definition of the system factor.

So long as the condition concerning the compositions of
the analyzed and reference samples can be met at all, it will
be met sufficiently well only if the concentrations of the components to be determined and of the standard and reference compounds are very low. In this case, slight variations in the
concentrations will not cause significant changes in their distribution ratios; the latter will be determined by the composition of the parent material and the extrahent. With reference
to Eqs. (7-25), (7-26), and (7-31)-(7-33), the above concepts
can be expressed mathematically as follows:

a. *Absolute-Calibration Technique*. Let us rewrite Eq.
(7-25) as

$$q_i = \frac{A_{\underline{i}e}}{A_s} \frac{v_{(s)}}{v_{(i)\underline{e}}} q_s \left[\varphi \left(f_i^W, f_s^W, \frac{V_{(i)e}}{V_{(i)}}, f_{\underline{i}\underline{p}\underline{e}} \right) \right] \qquad (7\text{-}35)$$

where φ is a factor comprising the quantities (in the parentheses) that are constants or should be kept the same in both the
calibration and the analytical procedures. Designating the
variables referring to the calibration procedure by an aster-
isk, we can write analogously

$$q_i^* = \frac{A_{\underline{i}e}^* \; v_{(s)}^*}{A_s^* \; v_{(i)\underline{e}}^*} \; q_s^* \left[\varphi \left(f_i^W, \; f_s^W, \; \frac{V_{(i)\underline{e}}}{V_{(i)}}, \; f_{i\underline{pe}} \right) \right] \tag{7-36}$$

and combining Eqs. (7-35) and (7-36) gives

$$q_i = \frac{A_{\underline{i}e} \; A_s^*}{A_{\underline{i}e}^* \; A_s} \; \frac{v_{(s)}}{v_{(s)}^*} \; \frac{v_{(i)\underline{e}}^*}{v_{(i)\underline{e}}} \; \frac{q_s}{q_s^*} \; q_i^* \tag{7-37}$$

By applying the same presuppositions and notations also to the other techniques, we obtain the following:

b.   *Internal-Standard Technique*

$$q_i = \frac{W_s}{W_s^*} \; \frac{v_{(i)\underline{e}}^{O*}}{v_{(i)\underline{e}}^O} \; \frac{A_{\underline{i}e}'}{A_{\underline{i}e}'^*} \; \frac{A_s'^*}{A_s'} \; q_i^* \tag{7-38}$$

c.   *Standard-Addition Technique: Measurement of the Sample Charge*

$$q_i = \frac{W_s}{W_s^*} \; \frac{v_{(i)\underline{e}}^{O*}}{v_{(i)\underline{e}}^O} \; \frac{\dfrac{A_{\underline{i}se}'^* \; v_{(i)\underline{e}}^*}{A_{\underline{i}e}^* \; v_{(i)\underline{e}}'^*} \left[1 + \dfrac{v_s^*}{v_{(i)\underline{e}}^{O*}} \right] - 1}{\dfrac{A_{\underline{i}se}' \; v_{(i)\underline{e}}}{A_{\underline{i}e} \; v_{(i)\underline{e}}'} \left[1 + \dfrac{v_s}{v_{(i)\underline{e}}^O} \right] - 1} \; q_i^* \tag{7-39}$$

d.   *Standard-Addition Technique: Use of a Reference Compound*

$$q_i = \frac{W_s}{W_s^*} \; \frac{v_{(i)\underline{e}}^{O*}}{v_{(i)\underline{e}}^O} \; \frac{(A_{\underline{i}se}'^* / A_{\underline{i}e}^*)(A_{\underline{pe}}^* / A_{\underline{pe}}'^*) - 1}{(A_{\underline{i}se}' / A_{\underline{i}e})(A_{\underline{pe}} / A_{\underline{pe}}') - 1} \tag{7-40}$$

As for the internal-normalization technique, elimination
or determination of all the system factors would render it
necessary to prepare either a mixture in which all the components involved in the normalization procedure would have
the same proportions as in the sample analyzed, or as many
mixtures as is the number of the components, with various proportions of the latter, and to solve the corresponding set of
equations. Hence, the normalization technique is hardly practicable in connection with the procedures involving sample
extraction.

It happens very frequently that the composition of the
parent material, i.e., that constituted by the major components, is unknown, so that the model mixture of the composition
required cannot be prepared. Regrettably, this is always the
case with biological materials, such as diverse tissues, blood,
urine, etc. In such cases, the above-described alternatives
to the individual techniques are inapplicable. Fortunately,
the standard-addition technique affords alternatives which
do not suffer from this qualification.

2. *Use of the Standard-Addition Technique*

The standard-addition technique has a distinct advantage
over all the other techniques in combination with sample
extraction. This advantage rests with the fact that the
addition of a relatively small amount of the component under
determination to the sample will not alter appreciably the
composition of the latter, so that the chemical potentials
of the components to be determined in the material with and
without the added standard will be virtually the same.

The standard-addition technique combined with extraction
offers two alternatives. Namely, it is possible to work either
with two separate samples, the original one and the sample-
standard mixture (Alternative I), or with a single sample

(Alternative II). In the latter case, a sample of the extract
of the original sample is analyzed first, whereupon the standard
is added to the sample-extract system and a sample of the extract is analyzed again. With both alternatives, the systems
with the initial and the standardized samples must be kept at
the same temperature.

Alternative I. It follows from the above discussion that,
when proceeding in the same manner as described for the ordinary
(nonextraction) standard-addition technique while incorporating
the extraction step and chromatographing the extracts of the
sample and sample-standard mixture instead of the materials as
such, the relations derived for the ordinary standard-addition
technique will apply for the procedure involving extraction as
well. The only and obvious qualifications for this to be true
are that the systems with and without the added standard must
be kept at the same temperature and that the ratios of the original sample to the extrahent and the sample-standard mixture to
the extrahent must be the same. This assertion can be proved
as follows:

The concentration of component i in the original sample
and in the sample-standard mixture, q_i and q_i', are given by

$$q_i = \frac{W_i}{V_{(i)}} = \frac{W_{ie} f_{ipe}}{V_{(i)}} \qquad\qquad (7\text{-}41)$$

and

$$q_i' = \frac{W_i + W_s}{V_{(i)} + V_s} = \frac{(W_{ie} + W_{se}) f_{ipe}'}{V_{(i)} + V_s} \qquad\qquad (7\text{-}42)$$

where f_{ipe}' refers to the sample-standard mixture. So long as
the amount of the standard substance (together with its solvent,
if any) is negligible in comparison with the sample amount, the

relation $f_{ipe} = f'_{ipe}$ applies to a good approximation at
constant temperature, and we have

$$\frac{q'_i}{q_i} = \frac{W_{ie} + W_{se}}{W_{ie}} \frac{V_{(i)}}{V_{(i)} + V_s} \tag{7-43}$$

The concentrations of component i in the extracts of the sample
and the sample-standard mixture, q_{ie} and q'_{ie}, are

$$q_{ie} = \frac{W_{ie}}{V_{(i)e}} \tag{7-44}$$

and

$$q'_{ie} = \frac{W_{ie} + W_{se}}{V'_{(i)e}} \tag{7-45}$$

where $V'_{(i)e}$ is the volume of the extract of the sample-standard
mixture. Hence,

$$\frac{q'_{ie}}{q_{ie}} = \frac{W_{ie} + W_{se}}{W_{ie}} \frac{V_{(i)e}}{V'_{(i)e}} \tag{7-46}$$

and combining Eqs. (7-46) and (7-43) gives

$$\frac{q'_i}{q_i} = \frac{q'_{ie}}{q_{ie}} \frac{V'_{(i)e}}{V_{(i)e}} \frac{V_{(i)}}{V_{(i)} + V_s} \tag{7-47}$$

The requirement that the ratios $V_{(i)p}/V_{(i)e}$ and $V'_{(i)p}/V'_{(i)e}$
must be the same (a necessary condition for $f_{ipe} = f'_{ipe}$)
implies that the ratios $V_{(i)}/V_{(i)e}$ and $(V_{(i)} + V_s)/V'_{(i)e}$ are
also the same, and Eq. (7-47) can be simplified to

$$\frac{q'_i}{q_i} = \frac{q'_{ie}}{q_{ie}} \tag{7-48}$$

The weights w_{ie} and w'_{ie} of component i contained in the volume charges $v_{(i)e}$ and $v'_{(i)e}$ of the extracts of sample and sample-standard mixture are related to the respective peak areas A_{ie} and A'_{ise} by

$$w'_{ie} = q'_{ie} v'_{(i)e} = CA'_{ise} \tag{7-49}$$

$$w_{ie} = q_{ie} v_{(i)e} = CA_{ie} \tag{7-50}$$

so that

$$\frac{q'_{ie}}{q_{ie}} = \frac{A'_{ise}}{A_{ie}} \frac{v_{(i)e}}{v'_{(i)e}} \tag{7-51}$$

For q'_i/q_i, it follows from Eqs. (7-41) and (7-42) that

$$\frac{q'_i}{q_i} = \left(1 + \frac{W_s}{W_i}\right) \frac{V_{(i)}}{V_{(i)} + V_s} \tag{7-52}$$

Hence, solving Eqs. (7-48), (7-51), and (7-52) for W_i and replacing the latter by $q_i V_{(i)}$ gives

$$q_i = \frac{W_s}{V_{(i)}} \left[\frac{A'_{ise}}{A_{ie}} \frac{v_{(i)e}}{v'_{(i)e}} \left(1 + \frac{V_s}{V_{(i)}}\right) - 1 \right]^{-1} \tag{7-53}$$

With regard to the qualifications discussed above, V_s will always be negligible in comparison with $V_{(i)}$, so that the correction term $\{[V_s/V_{(i)}] + 1\}$ has little bearing on this case.

For the standard-addition technique with an auxiliary reference substance, applied under the same circumstances as discussed above, it can directly be written

$$q_i = \frac{W_s}{V_{(i)}} \frac{1}{(A'_{ise}/A_{ie})(A_{pe}/A'_{pe}) - 1} \tag{7-54}$$

provided that the reference compound p was present (or added)
in the sample prior to its mixing with the standard and adding
the extrahent.

   Alternative II. This alternative is applicable only if a
negligible volume of the extract of the original sample, $v_{(i)\underline{e}}$,
as compared with the overall volume of the extract, $V_{(i)\underline{e}}$, is
taken for analysis. Only in this case would the volumes $V_{(i)\underline{e}}$
and $V'_{(i)\underline{e}}$ as well as the composition of the system before and
after the sample withdrawal and addition of the standard
remain practically unchanged. For the first step, i.e., the
analysis of the extract of the original sample, we have [cf.
Eq. (7-23)]

$$W_i = W_{i\underline{e}} \left(\frac{K_{-i\underline{p}\underline{e}} V_{(i)\underline{p}}}{V_{(i)\underline{e}}} + 1 \right) \tag{7-55}$$

After taking a volume $v_{(i)\underline{e}}$ out of the extract, representing a
weight amount $w_{i\underline{e}} = q_{i\underline{e}} v_{(i)\underline{e}}$ of component i, and adding a weight
amount W_s of the standard into the system, the total weight
amount W'_i of component i in the system will be given by

$$W'_i = W_i + W_s - w_{i\underline{e}} \tag{7-56}$$

which can further be written out as

$$W'_i = W_{i\underline{p}} + W_{i\underline{e}} + W_{s\underline{p}} + W_{s\underline{e}} - w_{i\underline{e}} \tag{7-57}$$

Provided the overall composition of the system is not altered
appreciably by taking out the sample of the extract and adding
the standard, the quantities referring to the parent solution
are colligated with those referring to the extract through the
partition coefficient $K_{-i\underline{p}\underline{e}}$, and, if $v_{(i)\underline{e}} \ll V_{(i)\underline{e}}$, Eq. (7-57)
can be rewritten to read

$$w'_i = \left(w_{i\underline{e}} + w_{s\underline{e}}\right) \left[\frac{\dfrac{K_{i\underline{p}\underline{e}}V_{(i)\underline{p}}}{V_{(i)\underline{e}}} + 1\right] - w_{i\underline{e}} \tag{7-58}$$

If $v_{(i)\underline{e}} \ll V_{(i)\underline{e}}$, then also $w_{i\underline{e}} \ll W_{i\underline{e}} + W_{s\underline{e}}$, and $w_{i\underline{e}}$ need not be considered. Hence, with regard to Eqs. (7-58), (7-55), and (7-56), we can write

$$\frac{w'_i}{w_i} = \frac{W_{i\underline{e}} + W_{s\underline{e}}}{W_{i\underline{e}}} = \frac{W_i + W_s}{W_i} = \frac{W'_{i\underline{e}}}{W_{i\underline{e}}} \tag{7-59}$$

from which it follows that

$$W_i = \frac{W_s}{(W'_{i\underline{e}}/W_{i\underline{e}}) - 1} \tag{7-60}$$

Expressed in terms of concentration, Eq. (7-60) reads

$$q_i = \frac{W_s}{V_{(i)}} \frac{1}{(q'_{i\underline{e}}/q_{i\underline{e}})(V'_{(i)\underline{e}}/V_{(i)\underline{e}}) - 1} \tag{7-61}$$

The ratio $V'_{(i)\underline{e}}/V_{(i)\underline{e}}$ must approach unity, and $q'_{i\underline{e}}/q_{i\underline{e}}$ is given [cf. Eq. (7-51)] by $A'_{is\underline{e}}v_{(i)\underline{e}}/A_{i\underline{e}}v'_{(i)\underline{e}}$ at constant chromatographic conditions, so that we have

$$q_i = \frac{W_s}{V_{(i)}} \frac{1}{(A'_{is\underline{e}}/A_{i\underline{e}})(v_{(i)\underline{e}}/v'_{(i)\underline{e}}) - 1} \tag{7-62}$$

Comparison of Eqs. (7-62) and (7-53) shows that the single-sample standard-addition technique is essentially equivalent to the more customary double-sample procedure under the above-outlined conditions.

3. Double-Extraction Technique

The system factor can also be eliminated by subsequently carrying out two extractions of the given sample and determining the contents of component i in the two extracts. This possibility is of course again qualified by the requirements that the concentration of the component to be determined must be very low, that the other (major) components do not pass appreciably into the extrahent phase, that the volume ratios of the parent solution to the extract be the same in both extraction steps, and that the system is kept at a constant temperature. Thus, designating the symbols referring to the first and second extraction by subscripts 0 and 1, we can write, with reference to Eq. (7-23),

$$W_i = W_{ie0} f_{ipe} \qquad (7\text{-}63)$$

$$W_{i1} = W_{ie1} f_{ipe} \qquad (7\text{-}64)$$

Provided that the extract from the first extraction is completely separated from the parent solution, the difference between W_i and W_{i1} is just the weight of component i present in the first extract, i.e.,

$$W_{i1} = W_i - W_{ie0} \qquad (7\text{-}65)$$

and combining Eqs. (7-63)-(7-65) gives

$$W_i = \frac{W_{ie0}}{1 - (W_{ie1}/W_{ie0})} \qquad (7\text{-}66)$$

When expressing the absolute weights by means of the respective concentrations, Eq. (7-66) becomes

$$q_i = \frac{q_{ie0} V_{(i)e0}/V_{(i)}}{1 - (q_{ie1} V_{(i)e1}/q_{ie0} V_{(i)e0})} \qquad (7\text{-}67)$$

The above-stated qualifying requirements imply that the volumes $V_{(i)e1}$ and $V_{(i)e0}$ will be approximately the same, so that their ratio in Eq. (7-67) will approach unity, unless the volumes of the initial sample and the parent solution taken for the second extraction are different. The values of q_{ie1} and q_{ie0} can be determined by any of the conventional techniques of quantitative GC.

C. Condensing the Extract

It may be expedient in many cases that the concentration of the component to be determined in the extract be increased by evaporating a part of the extrahent. In quantitative analysis, this obviously has to be carried out in a defined manner. Let us suppose that volume $V^o_{(i)e}$ taken from the original extract is condensed to volume $V''_{(i)e}$. Provided that the absolute amount of the component under determination remains unchanged, the concentration of this component in the condensed extract, q''_{ie}, is given by

$$q''_{ie} = \frac{q_{ie} V^o_{(i)e}}{V''_{(i)e}} \tag{7-68}$$

where q_{ie} is the initial concentration.

1. Direct Analysis of the Condensed Extract

When applying the concentration step in procedures without elimination of the system factor [cf. Eqs. (7-25), (7-26), and (7-31)-(7-34)], we can write for the individual techniques:

a. Absolute-Calibration Technique

$$q_i = \frac{A_{ie}}{A_s} \frac{f^W_i}{f^W_s} \frac{V_{(s)}}{V''_{(i)e}} \frac{V''_{(i)e}}{V^o_{(i)e}} \frac{V_{(i)e}}{V_{(i)}} q_s f_{ipe} \tag{7-69}$$

where $v''_{(i)e}$ is the volume of the concentrated extract introduced into the gas chromatograph.

   b. Internal-Standard Technique. If the condensing procedure is applied after the standard substance has been added
to the extract, Eq. (7-26) holds without change. If, on the
other hand, volume $v^o_{(i)e}$ of the initial extract is first condensed to volume $V''_{(i)e}$ and volume $V''^o_{(i)e}$ of the condensed
extract mixed with the standard, the following equation
applies:

$$q_i = \frac{W_s}{V''^o_{(i)e}} \frac{V''_{(i)e}}{V^o_{(i)e}} \frac{V_{(i)e}}{V_{(i)}} \frac{A'_{ie}}{A'_s} \frac{f^W_i}{f^W_s} f_{ipe}. \qquad (7\text{-}70)$$

If $V''^o_{(i)e} = V''_{(i)e}$, i.e., if the entire volume of the condensed
extract is processed, Eq. (7-70) becomes identical with Eq.
(7-26) as well.

   In the version described by Eq. (7-31), condensing the
material to be introduced into the gas chromatograph has no
effect on the calculation procedure, i.e., Eq. (7-31) is
readily applicable.

   c. Standard-Addition Technique. Provided that both chromatographic runs, i.e., those with and without the addition of
the standard, are performed with the condensed extract, the
term $W_s V_{(i)e}/V^o_{(i)e}V_{(i)}$ in Eqs. (7-32) and (7-33) will be

replaced by $W_s V''_{(i)e}V_{(i)e}/V''^o_{(i)e}V^o_{(i)e}V_{(i)}$ and the term

$[v_{(i)e}/v'_{(i)e}]\{1 + [V_s/V_{(i)e}]\}$ in Eq. (7-32) by

$[v''_{(i)e}/v'''_{(i)e}]\{1 + [V_s/V''^o_{(i)e}]\}$.

   The internal-normalization method as described by Eq.
(7-34) is unaffected by condensing the extract.

2. *Use of Reference Model Mixtures*

For the methods involving elimination of the system factor through comparison with a model system [cf. Eqs. (7-37)-(7-40)], independent condensation of the extracts of the analyzed and reference systems renders

$$q_i = \frac{A_{ie}}{A_{ie}^*} \; \frac{A_s^*}{A_s} \; \frac{v_{(s)}}{v_{(s)}^*} \; \frac{v''^*_{(i)e}}{v''_{(i)e}} \; \frac{v''_{(i)e}}{v''^*_{(i)e}} \; \frac{v^{O*}_{(i)e}}{v^{O}_{(i)e}} \; \frac{q_s}{q_s^*} \; q_i^* \tag{7-71}$$

$$q_i = \frac{W_s}{W_s^*} \; \frac{v''^{O*}_{(i)e}}{v''^{O}_{(i)e}} \; \frac{v''_{(i)e}}{v''^*_{(i)e}} \; \frac{v^{O*}_{(i)e}}{v^{O}_{(i)e}} \; \frac{A'_{ie}}{A'_{ie}} \; \frac{A'^*_s}{A'_s} \; q_i^* \tag{7-72}$$

for the absolute-calibration and internal-standard techniques, and

$$q_i = \frac{W_s}{W_s^*} \; \frac{v''^{O*}_{(i)e}}{v''^{O}_{(i)e}} \; \frac{v''_{(i)e}}{v''^*_{(i)e}} \; \frac{v^{O*}_{(i)e}}{v^{O}_{(i)e}} \; \frac{\theta^*}{\theta} \; q_i^* \tag{7-73}$$

where

$$\theta^* = \frac{A'^*_{ise}}{A_{ie}^*} \; \frac{v''^*_{(i)e}}{v'''^*_{(i)e}} \left(1 + \frac{v_s^*}{v''^{O*}_{(i)e}} \right) - 1$$

$$\theta = \frac{A'_{ise}}{A_{ie}} \; \frac{v''_{(i)e}}{v'''_{(i)e}} \left(1 + \frac{v_s}{v''^{O}_{(i)e}} \right) - 1$$

and

$$q_i = \frac{W_s}{W_s^*} \frac{V''^{O*}_{(i)\underline{e}}}{V''^{O}_{(i)\underline{e}}} \frac{V''_{(i)\underline{e}}}{V''^{*}_{(i)\underline{e}}} \frac{V^{O*}_{(i)\underline{e}}}{V^{O}_{(i)\underline{e}}} \frac{\Theta*}{\Theta} q_i^* \qquad (7\text{-}74)$$

where

$$\Theta* = \frac{A'^*_{is\underline{e}}}{A^*_{i\underline{e}}} \frac{A^*_{p\underline{e}}}{A'^*_{p\underline{e}}} - 1$$

$$\Theta = \frac{A'_{is\underline{e}}}{A_{i\underline{e}}} \frac{A_{p\underline{e}}}{A'_{p\underline{e}}} - 1$$

for the two standard-addition techniques. Equations (7-73)
and (7-74) apply for the case in which the condensed extracts
of the analyzed and reference materials are processed.

Owing to the qualifications discussed above, it is expedi-
ent that the ratios $V''_{(i)\underline{e}}/V^{O}_{(i)\underline{e}}$ and $V''^{*}_{(i)\underline{e}}/V^{O*}_{(i)\underline{e}}$ be kept invariant and Eqs. (7-71)-(7-74) simplified accordingly.

3. Use of the Standard-Addition Technique

Alternative I. Let us consider the case in which samples
[volumes $V^{O'}_{(i)\underline{e}}$ and $V^{O}_{(i)\underline{e}}$] of the extracts of the systems with
and without the added standard are independently condensed to
volumes $V'''_{(i)\underline{e}}$ and $V''_{(i)\underline{e}}$, from which charges $v'''_{(i)\underline{e}}$ and $v''_{(i)\underline{e}}$
are injected into the gas chromatograph, respectively. It can
be shown that in this case

$$\frac{q'_{i\underline{e}}}{q_{i\underline{e}}} = \frac{A'_{is\underline{e}}}{A_{i\underline{e}}} \frac{v''_{(i)\underline{e}}}{v'''_{(i)\underline{e}}} \frac{V^{O}_{(i)\underline{e}}/V''_{(i)\underline{e}}}{V^{O'}_{(i)\underline{e}}/V'''_{(i)\underline{e}}} \qquad (7\text{-}75)$$

where

$$\frac{v''_{(i)\underline{e}}}{v'''_{(i)\underline{e}}} \frac{v^o_{(i)\underline{e}}/v''_{(i)\underline{e}}}{v^{o'}_{(i)\underline{e}}/v'''_{(i)\underline{e}}} = \frac{A_{p\underline{e}}}{A'_{p\underline{e}}} \frac{v^o_{(i)\underline{e}}}{v^o_{(i)\underline{e}} + V_s} \qquad (7\text{-}76)$$

Hence,

$$q_i = \frac{W_s}{V_{(i)}} \left[\frac{A'_{is\underline{e}}}{A_{i\underline{e}}} \frac{v''_{(i)\underline{e}}}{v'''_{(i)\underline{e}}} \frac{v^o_{(i)\underline{e}}/v''_{(i)\underline{e}}}{v^{o'}_{(i)\underline{e}}/v'''_{(i)\underline{e}}} \left(1 + \frac{V_s}{v^o_{(i)\underline{e}}} \right) - 1 \right]^{-1} \qquad (7\text{-}77)$$

and

$$q_i = \frac{W_s}{V_{(i)}} \left(\frac{A'_{is\underline{e}}}{A_{i\underline{e}}} \frac{A_{p\underline{e}}}{A'_{p\underline{e}}} - 1 \right)^{-1} \qquad (7\text{-}78)$$

The derivation of Eqs. (7-77) and (7-78) involves the pre-supposition that the absolute contents of the components i and p in the sample of the extracts remain unchanged during the evaporation of the extrahent. As this often is not the case, it is advisable to keep the ratios $v^o_{(i)\underline{e}}/v''_{(i)\underline{e}}$ and $v^{o'}_{(i)\underline{e}}/v'''_{(i)\underline{e}}$ as well as the entire concentration procedure (absolute volumes, temperature, rate of extrahent evaporation) the same in both steps.

With the single-sample standard-addition technique, the requirement that the first sample of the extract be taken as small as possible makes the use of a condensed extract impracticable.

Alternative II. In the double-extraction technique, samples $v^o_{(i)\underline{e}0}$ and $v^o_{(i)\underline{e}1}$ of the first and the second extract can be taken and condensed to volumes $V''_{(i)\underline{e}0}$ and $V''_{(i)\underline{e}1}$, respectively. As $W_{i\underline{e}0} = W^o_{i\underline{e}0} V_{(i)\underline{e}0}/v^o_{(i)\underline{e}0}$ and $W_{i\underline{e}1} = W^o_{i\underline{e}1} V_{(i)\underline{e}1}/v^o_{(i)\underline{e}1}$, where $W^o_{i\underline{e}0}$ and $W^o_{i\underline{e}1}$ are the absolute weight amounts of

component i in the samples of the extracts, and W_{ie0}^{o} =
$q_{ie0}^{''}V_{(i)e0}^{''}$ and $W_{ie1}^{o} = q_{ie1}^{''}V_{(i)e1}^{''}$, we can write, with regard to
Eq. (7-66),

$$q_i = \frac{q_{ie0}^{''}V_{(i)e0}^{''}V_{(i)e0}^{V}/V_{(i)e0}^{o}V_{(i)}^{V}}{1 - \dfrac{q_{ie1}^{''}V_{(i)e1}^{''}V_{(i)e1}^{V}/V_{(i)e1}^{o}}{q_{ie0}^{''}V_{(i)e0}^{''}V_{(i)e0}^{V}/V_{(i)e0}^{o}}} \qquad (7\text{-}79)$$

where $q_{ie0}^{''}$ and $q_{ie1}^{''}$ are the concentrations of component i in
the condensed samples of the extracts. In this technique,
the volumes $V_{(i)e0}^{o}$ and $V_{(i)e1}^{o}$ should be the same and processed
in the same manner.

D. *Multiple Extraction*

By extracting repeatedly a given volume of the material
to be analyzed with volumes of fresh extrahent while separating and collecting the individual extracts, it is possible
to achieve a practically complete transfer of the components
of interest from the parent material into the joined extracts
after some number of extraction steps. The volume of joined
extracts can then be condensed by evaporating the extrahent,
and the content of component i in the condensate determined
by any of the conventional techniques of quantitative gas
chromatography.

Let us suppose that volume $V_{(i)}$ of the material to be
analyzed is extracted with equal volumes V_e of the extrahent,
the number of the extraction steps being n. The weight amount
of component i, $W_{ip,n}$, left in the parent solution after the
n^{th} extraction is

$$W_{i\underline{p},n} = \frac{W_i}{\left[\left(V_{(i)\underline{e}}/V_{(i)\underline{p}}K_{\underline{i}\underline{p}\underline{e}}\right) + 1\right]^n} \tag{7-80}$$

where $V_{(i)\underline{e}}$ and $V_{(i)\underline{p}}$ are approximately equal to $V_{(i)}$. Hence, we can write for W_i

$$W_i = \sum_n W_{i\underline{e}} + W_{i\underline{p},n} \tag{7-81}$$

where $\sum_n W_{i\underline{e}}$ is the sum of the amounts of component i present in the extracts. The mean concentration of component i in the collected extracts, $\bar{q}_{i\underline{e}}$, is

$$\bar{q}_{i\underline{e}} = \frac{\sum_n W_{i\underline{e}}}{nV_{(i)\underline{e}}} \tag{7-82}$$

As $q_i = W_i/V_{(i)}$, we can write, with regard to Eqs. (7-81) and (7-82),

$$q_i = \frac{n\bar{q}_{i\underline{e}}V_{(i)\underline{e}} + W_{i\underline{p},n}}{V_{(i)}} \tag{7-83}$$

If the number of extractions is sufficiently large, the term $W_{i\underline{p},n}$ in Eq. (7-83) becomes negligible in comparison with $n\bar{q}_{i\underline{e}}V_{(i)\underline{e}}$, i.e., component i is practically completely transferred into the extract of volume $nV_{(i)\underline{e}}$. If volume $nV_{(i)\underline{e}}$ is condensed to volume $V''_{(i)\underline{e}}$, the concentration of component i will increase from $\bar{q}_{i\underline{e}}$ to $q''_{i\underline{e}}$; provided no losses of component i occur during condensation of the extract,

$$\bar{q}_{i\underline{e}} = \frac{q''_{i\underline{e}}V''_{(i)\underline{e}}}{nV_{(i)\underline{e}}} \tag{7-84}$$

and we can write

$$q_i = \frac{q''_{i\underline{e}} V''_{(i)\underline{e}}}{V_{(i)}} \tag{7-85}$$

where $q''_{i\underline{e}}$ can again be determined by any quantitative GC technique. Equation (7-85) indicates that fairly high degree of
quantitative accumulation of component i isolated from excess
ballast material can be achieved by the multiple-extraction
procedure.

III.   HEAD-SPACE GAS ANALYSIS

The interpretation of the term "head-space gas analysis"
is rather facultative. There are roughly two conceptions of
this method; it is understood either as the analysis of the
gaseous phase the composition of which is equilibrated with
that of the coexisting liquid phase in a closed system, or as
the determination of volatile components stripped with an inert
gas out of liquid material. In both of these alternatives,
very often diverse trapping tubes are employed for sampling the
head-space gas. In the latter alternative, this kind of sampling is usually a matter of necessity as the components to be
determined are diluted in a large volume of the stripping gas
and have to be concentrated. However, the use of a trapping
tube may also be of advantage in the former alternative, because fairly large volumes of the head-space gas can be drawn
through the tube, thus achieving a sufficiently high degree
of accumulation of the components to be determined while avoid-
ing the problems incidental to injecting large sample volumes
into the gas chromatograph. As for the sampling proper and
further treatment of the sampling tube, the same rules apply
as discussed in Section I of this chapter.

The use of a sampling tube allows an interesting version
of head-space gas analysis to be performed, in which the

components of interest are stripped out in a closed loop with
the trapping tube connected in series, i.e., the stripping is
carried out by recycling the same head-space gas, which is
cleared gradually of the components to be determined by trapping them in the tube [111, 112].

In spite of the extensive use of GC head-space gas analysis in a number of important areas, the exact quantitation,
though feasible, is omitted, the interpretation of the results
being usually restricted to mere fingerprint comparisons.

A. Sampling from a Closed Gas-Liquid System

In this case, the analytical task is to determine the total
contents of certain components in the entire two-phase system
by analyzing just a sample taken from the gaseous phase enclosed
above the liquid. The equilibrium between both phases is characterized by a partition coefficient K_{LG}. For a component i,
the partition coefficient K_{iLG} is

$$K_{iLG} = \frac{q_{iL}}{q_{iG}} = \frac{W_{iL}/V_L}{W_{iG}/V_G} \tag{7-86}$$

where V_G and V_L are the volumes of the gaseous and liquid phases
in the system and W_{iL} and W_{iG} are the weight amounts of component i contained in these phases, respectively. The total amount
of component i in the system, W_i, is obviously given by

$$W_i = W_{iL} + W_{iG} \tag{7-87}$$

which, on combining with Eq. (7-86), gives

$$W_i = W_{iG} \left(\frac{K_{iLG} V_L}{V_G} + 1 \right) \tag{7-88}$$

Equation (7-88) shows that this alternative again involves the problem of a system factor [$(K_{iLG}V_L/V_G) + 1$], and it can easily be deduced that this problem is quite analogous to that discussed in the section on liquid extraction. This analogy is obviously given by the analogous nature of both problems; in both cases, the analytical task is the determination of the overall content of a component of an equilibrated two-phase system by analyzing a sample of merely one of the phases. Hence, it follows that all the procedures discussed with the liquid-extraction method can readily be applied in this version of head-space gas analysis as well. In this subsection, only the methods which involve the elimination of the system factor through use of the standard-addition and double-sampling techniques are discussed in detail, as only these appear to be of great practical importance.

Although the situations in the analysis of liquid-liquid and gas-liquid systems are basically the same, there are some important differences in the properties of these two systems, having certain analytical implications. For instance, the possibility of controlling the value of the system factor by the choice of the extrahent in liquid-liquid extraction is evidently out of the question in the case of gas-liquid systems. On the other hand, unlike with liquid-liquid systems, the equilibrium compositions of the phases in gas-liquid systems are very sensitive to temperature; upon increasing the temperature of a gas-liquid system analyzed for the content of component i, the value of K_{iLG} decreases sharply and the system factor approaches to unity. Hence, at a sufficiently high temperature, $W_i \doteq W_{iG}$, especially if $V_G \gg V_L$ [cf. Eq. (7-88)]. A typical example of the utilization of this phenomenon is the determination of alcohol in blood by the method of Machata [113].

In this context it may be recalled that, with all the
techniques involving the elimination of the system factor, the
original and the standardized (or calibrating) systems must be
kept at the same temperature when the respective samples are
taken for analysis. With gas-liquid systems, this requirement
is of much greater importance than with liquid-liquid systems.

The effect of temperature on the equilibrium distribution
of component i in a gas-liquid system can be illustrated as
follows: The partial pressure P_i of component i in the gaseous
phase of a closed gas-liquid system at equilibrium is given by

$$P_i = Px_{iG} = \gamma_{iL}P_i^o x_{iL} \qquad (7\text{-}89)$$

where P is the overall pressure in the system, P_i^o is the saturation vapor pressure of pure component i at the temperature
of the system, x_{iG} and x_{iL} are the mole fractions of component
i in the gas and in the liquid phases, and γ_{iL} is the Raoult-
law activity coefficient of component i in the liquid phase.
Employing the perfect-gas state equation, P_i can further be
expressed as

$$P_i = \frac{N_{iG}RT}{V_G} = \frac{W_{iG}RT}{V_G M_i} \qquad (7\text{-}90)$$

where N_{iG} is the number of moles of component i occurring in
the gaseous phase, R is the gas constant, M_i is the molecular
weight of component i, and T is the absolute temperature of
the system, W_{iG} and V_G being again the weight of component i
in the gaseous phase and the volume of the latter, respectively.
Equations (7-89) and (7-90), solved for W_{iG}, give

$$W_{iG} = \frac{\gamma_{iL}P_i^o x_{iL}M_i V_G}{RT} \qquad (7\text{-}91)$$

The quantity x_{iL} is given by

$$x_{iL} = \frac{N_{iL}}{N_L} = \frac{W_{iL}}{W_L}\frac{M'_L}{M_i} = \frac{W_{iL}}{V_L d_L}\frac{M'_L}{M_i} \tag{7-92}$$

where N_{iL} and W_{iL} are the mole number and weight of component
i present in the liquid phase, and W_L, V_L, d_L, and M'_L are the
weight, volume, density, and mean molecular weight of the
liquid phase, respectively. Combining Eqs. (7-91) and (7-92)
and substituting q_{iG} and q_{iL} for W_{iG}/V_G and W_{iL}/V_L, respectively,
we obtain, after some rearrangement,

$$\frac{q_{iL}}{q_{iG}} = K_{iLG} = \frac{RTd_L}{\gamma_{iL}P_i^o M'_L} \tag{7-93}$$

from which the role of temperature is readily apparent, viz.,
the quantity P_i^o is an exponential function of temperature.
Hence, the requirement that the temperature be invariant when
eliminating the system factor can hardly be overemphasized
with gas-liquid systems.

The same considerations applied to liquid-liquid (extraction) system would lead to

$$\frac{q_{ip}}{q_{ie}} = K_{ipe} = \frac{\gamma_{ie}M'_e d_p}{\gamma_{ip}M'_p d_e} \tag{7-94}$$

where the subscripts p and e refer to the parent solution and
the extract, respectively. This relation shows that the effect
of temperature is much less pronounced with liquid-liquid
systems.

There are also some distinctions when considering the
application of the single-sample standard-addition technique
and the double-extraction technique to head-space gas analysis.

In contrast to the situation in sampling the liquid extract
from a liquid-liquid system, the volume of the head space of a
gas-liquid system remains virtually unchanged even when fairly
large samples of the gaseous phase are taken. Further, upon
the withdrawal of a sample of the head-space gas, the partial
pressures of the components in the gas above their liquid solution are decreased, which produces temporarily a negative deviation from equilibrium in the gaseous phase and a transition
of the components from the liquid phase into the gaseous phase
in order to restore equilibrium. This is not the case with
liquid-liquid systems, as no concentration changes occur on
taking a sample of the extract. Hence, some modification of
the mass-balance equations quoted with the single-sample
extraction technique of standard-addition is necessary in
order to describe adequately the situation in the application
of this technique to head-space gas analysis.

1. *Single-Sample Standard-Addition Technique*

As the contents of component i in both the gaseous and
the liquid phase, $W_{i\underline{G}}$ and $W_{i\underline{L}}$, are subject to changes on
taking a sample of the gaseous phase $(w_i = w_{i\underline{L}} + w_{i\underline{G}})$,
equations

$$W_i' = W_i + W_s - w_i \tag{7-95}$$

$$W_i' = W_{i\underline{L}} + W_{i\underline{G}} + W_{s\underline{L}} + W_{s\underline{G}} - w_{i\underline{L}} - w_{i\underline{G}} \tag{7-96}$$

and

$$W_i' = (W_{i\underline{G}} + W_{s\underline{G}} - w_{i\underline{G}}) \left(\frac{K_{i\underline{L}\underline{G}} V_{\underline{L}}}{V_{\underline{G}}} + 1 \right) \tag{7-97}$$

apply in this case instead of Eqs. (7-56)-(7-58), respectively.
We can further write [cf. Eq. (7-59)]

$$\frac{W_{iG} + W_{sG} - w_{iG}}{W_{iG}} = \frac{W_i + W_s - w_i}{W_i} = \frac{W'_{iG}}{W_{iG}} \tag{7-98}$$

where the primed symbols again refer to the mixture of the sample and the added standard. Combining Eqs. (7-96)-(7-98) and solving for W_i gives

$$W_i = \frac{W_s - w_i}{(W'_{iG}/W_{iG}) - 1} \tag{7-99}$$

As the head-space gas volume is practically invariant in this alternative, the weight ratio W'_{iG}/W_{iG} is equal to the concentration ratio q'_{iG}/q_{iG}, and Eq. (7-99) can be rewritten as

$$W_i = \frac{W_s - w_i}{(A'_{isG}/A_{iG})(v_G/v'_G) - 1} \tag{7-100}$$

where v_G and v'_G are the volumes of the head-space gas samples taken for analysis before and after adding the standard, A_{iG} and A'_{isG} are the corresponding peak areas, W_i is the weight content of component i in the volume v_G, as determined in the first head-space gas sample, and W_s is the weight amount of the standard added to the system. When a sampling tube is employed, v_G and v'_G represent the volumes drawn through the tube.

In contrast to the corresponding liquid-extraction technique, it is not essential in this case that $v_G \ll V_G$ and $w_i \ll W_i + W_s$, but it is very important to arrange that the samples v_G and v'_G are withdrawn at the same rate, preferably as slowly as possible. Because the rate of reequilibration is rather low, different rates of the two samplings could completely disfigure the results.

2. Double-Sampling Technique

In comparison with the double-extraction technique, an
analogy applicable to head-space gas analysis is the double-
sampling technique. The corresponding relation [cf. Eq.
(7-66)] is written as

$$W_i = \frac{w_{iG0}}{1 - (w_{iG1}/w_{iG0})} \qquad (7\text{-}101)$$

where the subscripts 0 and 1 refer to the first and the second
samplings, and w_{iG0} is the amount of component i contained in
the sample volume taken in the first sampling step. As for the
rate of sampling, the same applies as has been stressed with the
preceding technique. In this case again, $W_{iG1}/W_{iG0} = q_{iG1}/q_{iG0} = (A_{iG1}/A_{iG0})(v_{G0}/v_{G1})$, where v_{G0} and v_{G1} are the sample volumes
taken from the head space in the first and the second sampling
steps, and A_{iG0} and A_{iG1} are the corresponding peak areas.

B. Analysis of Components Stripped Off the Sample

In this alternative [102, 114, 115], relatively large
volumes of the stripping gas have to be used, so that it is
usually inevitable that the components of interest become
concentrated in a trapping tube. The problem of quantitation
is rather trivial in this case. An exact quantitation is
feasible only if the entire content of the component to be
determined is stripped out of the sample and entrapped completely in the tube. The further treatment of the tube, i.e.,
the transfer of the concentrate into the gas chromatograph,
is again the same as described in Section I of this chapter.

The calibration can be carried out in two ways: a defined amount of a standard substance is either chromatographed
directly in a separate run or added to the sample analyzed.

The first method represents the conventional absolute calibration, and the calculation of the concentration of component i, q_i, in the original (liquid) sample is carried out by Eq. (7-22) where, in this case, $V_{(i)}$ is the volume of the liquid sample.

In the other method, it is possible to employ either the internal-standard or standard-addition technique. In the former case, the handling of the sample (apart from the stripping procedure and the manipulation with the tube) and the calculation of the results are the same as in the conventional internal-standard technique. When the standard-addition technique is employed, the stripping and trapping procedure eliminates completely the problem of volume corrections (factor $\{[V_s/V_{(i)}] - 1\}$); the results are calculated by

$$q_i = \frac{W_s}{V_{(i)}} \frac{1}{(A'_{is}/A_i) - (V'_{(i)}/V_{(i)})} \qquad (7\text{-}102)$$

where $V_{(i)}$ and $V'_{(i)}$ are the volumes of the original sample processed as such and after mixing it with the standard, respectively. Relation (7-102) can be derived as follows: The total amounts of component i in the original sample and in the sample-standard mixture, W_i and W'_i, are given by

$$W_i = q_i V_{(i)} \qquad (7\text{-}103)$$

and

$$W'_i = q_i V'_{(i)} + W_s \qquad (7\text{-}104)$$

As $W_i = CA_i$ and $W'_i = CA'_{is}$, it is possible to write

$$\frac{q_i V'_{(i)} + W_s}{q_i V_{(i)}} = \frac{A'_{is}}{A_i} \qquad (7\text{-}105)$$

which, solved for q_i, gives Eq. (7-102). The volume ratio $V'_{(i)}/V_{(i)}$ could advantageously be determined through use of an auxiliary reference substance, present in or added to the sample. The ratio $V'_{(i)}/V_{(i)}$ would then be given by the

corresponding ratio of the peak areas of the reference compound, A'_p/A_p. As the complete trapping of the components
stripped out of the sample is supposed to be achieved in both
runs, it is of no consequence here whether the same tube or
different tubes have been used for the sampling from the
original sample and from that enriched with the added
standard.

The application of the method of chromatographic equilibration in an on-line arrangement is obviously unfeasible
in connection with this alternative of head-space gas analysis, as the concentrations of the components in the gas
leaving the sample are subject to continuous changes during
the stripping procedure. However, the above method could be
applied in an off-line arrangement: The stripping gas must
be collected until the components to be determined have been
completely purged out of the sample, upon which, after the
concentrations of the components in the gas become equalized,
the gas can be analyzed by the equilibration method. In this
way we obtain the concentrations of the components in the
stripping gas; hence, it is necessary to know the volume of
this gas in order to calculate the corresponding concentrations in the liquid sample. Denoting the equalized concentration of component i in the stripping gas and the volume
of the latter q^+_{iG} and V^+_G, respectively, we obtain, for the
corresponding concentration q_i in the original liquid sample
of volume $V_{(i)}$,

$$q_i = \frac{q^+_{iG} V^+_G}{V_{(i)}} \tag{7-106}$$

Collecting and measuring the gas while keeping it free of
extraneous contaminants may, of course, present difficulties.

C.   Closed-Loop Stripping and Trapping
of the Components [111, 112]

This approach requires a more exacting instrumentation,
but, in return, it provides for straightforward on-line application of the equilibration-trapping method with a minimum
risk that the sampled gas will become contaminated with arti-
fact components. The recycling of the head-space gas is performed with the aid of a pump which repeatedly draws the gas
through the trapping tube and through or over the liquid sample until either the sample has completely been cleared of
the components to be determined, the latter being deposited
in the tube, or the whole liquid sample/head-space gas/samp-
ling-tube sorbent system has become equilibrated. Theoretically, the equilibrium should always be attained after the
gas has been recycling for a sufficiently long time, provided
that the components are sorbed reversibly in the sampling
tube. Hence, both alternatives of sampling and quantitation,
based either on the complete trapping or equilibration of the
components in the tube, are readily feasible with the closed-
loop procedure. In the former alternative, the calibration
and calculation of the results are exactly the same as in
simple open-loop stripping head-space gas analysis. The problems of quantitation in the equilibrium-trapping alternative
can be tackled as follows: Let the total weight amount of
component i, W_i, contained in a closed gas-liquid system be
the quantity to be determined, regardless of whether the whole
system or merely the liquid is the subject of analytical interest. After connecting this system into a closed loop with the
pump and trapping tube and recycling the gaseous phase until
equilibrium concentrations of component i are established in
all the three phases, i.e., in the recycling gas, the liquid
being passed through by the gas, and the packing of the trapping tube, the amount W_i is divided among the phases by

$$W_i = W_{iG}^{++} + W_{iL} + W_{iS}$$ (7-107)

where W_{iS} and W_{iG}^{++} are the amounts of component i present in the sorbent of the tube and in the void (gaseous) space of the loop, respectively. The amounts W_{iL} and W_{iS} are related to W_{iG}^{++} through the respective partition coefficients, K_{iLG} and K_{iSG}, by

$$\frac{W_{iL}}{V_L} = \frac{K_{iLG} W_{iG}^{++}}{V_G^{++}}$$ (7-108)

and

$$\frac{W_{iS}}{V_S} = \frac{K_{iSG} W_{iG}^{++}}{V_G^{++}}$$ (7-109)

where V_S is the volume of the sorbent contained in the sampling tube; V_G^{++} in Eqs. (7-108) and (7-109) is the gas volume within the entire loop, i.e., the sum of the gas volumes of the initial gas-liquid system, pump, sampling tube, and connections.

It is the amount of component i entrapped in the tube that will be subject to further analytical treatment. Hence, it is expedient to express the amount W_i as a function of W_{iS}. Such a relationship can be obtained by combining Eqs. (7-107)-(7-109) and solving for W_i/W_{iS} which gives, after rearrangement,

$$W_i = W_{iS}\left(\frac{K_{iLG}}{K_{iSG}} \frac{V_L}{V_S} + \frac{1}{K_{iSG}} \frac{V_G^{++}}{V_S} + 1\right)$$ (7-110)

It can easily be shown that the ratio K_{iLG}/K_{iSG} actually represents the partition coefficient K_{iLS}. It is interesting to note that, even if the sorbent in the tube were a liquid

miscible with the liquid of the system under analysis, the
quantity K_{iLS} would be amenable to experimental determination,
thanks to the separation of both liquids by the gaseous phase.

The calibration by the standard-addition technique consists in analyzing independently, under identical conditions,
two systems of the same volumes of the gaseous and liquid
phases, with and without the addition of the standard, respectively. The overall amount of component i deposited in the
tube, W_{iT}, is given by

$$W_{iT} = W_{iS} + W_{iGT} \tag{7-111}$$

where W_{iGT} is the amount of component i contained in the void
volume of the tube. For the case of the sample with the
added standard, we can analogously write

$$W'_{iT} = W'_{iS} + W'_{iGT} \tag{7-112}$$

As

$$W_i = W_{iS} f_{iLGS} \tag{7-113}$$

and

$$W'_i = W'_{iS} f_{iLGS} \tag{7-114}$$

where f_{iLGS} is a system factor given by the expression in the
parentheses on the right-hand side of Eq. (7-110), we can further write

$$\frac{W'_i}{W_i} = \frac{W_i + W_s}{W_i} = \frac{W'_{iT} - W'_{iGT}}{W_{iT} - W_{iGT}} \tag{7-115}$$

So long as W'_{iGT} and W_{iGT} can be neglected in comparison with
W'_{iT} and W_{iT}, which is nearly always the case, the solution of
Eq. (7-115) for W_i gives

$$W_i = \frac{W_s}{(W'_{i\underline{T}}/W_{i\underline{T}}) - 1} \qquad (7\text{-}116)$$

where $W'_{i\underline{T}}$ and $W_{i\underline{T}}$ are the amounts of component i contained in the sampling tube.

Determination of $W'_{i\underline{T}}$ and $W_{i\underline{T}}$ can be carried out by treatment of the sampling tube as described above (cf. Section I of this chapter).   If $W'_{i\underline{T}}$ and $W_{i\underline{T}}$ are determined independently, separate calibrations must be carried out and $W'_{i\underline{T}}$ and $W_{i\underline{T}}$ calculated by

$$W'_{i\underline{T}} = \frac{A'_{i\underline{T}}}{A_s} \frac{f^W_i}{f^W_s} q_s v(s) \qquad (7\text{-}117)$$

and

$$W_{i\underline{T}} = \frac{A_{i\underline{T}}}{A_s} \frac{f^W_i}{f^W_s} q_s v(s) \qquad (7\text{-}118)$$

However, so long as both determinations are carried out under identical chromatographic conditions, the ratio $W'_{i\underline{T}}/W_{i\underline{T}}$ is given directly by the ratio of the respective peak areas, $A'_{i\underline{T}}/A_{i\underline{T}}$.

The requirement that the system factor f_{iLGS} be eliminated in the calculation has associated with it some important qualifications: As the value of the system factor depends on the compositions and absolute volumes of all the three phases constituting the closed-loop system [cf. Eq. (7-110)], it is necessary that the volumes of the liquid and gaseous phases in both the original and standardized systems be the same and that the same sampling tube (the same kind and amount of the sorbent) be employed in both runs.   Further, it is necessary to maintain

all parts of the closed-loop system at a constant temperature
and to observe that the addition of the standard does not alter appreciably the composition of the system, in order to
keep the value of the system factor invariant. This will be
so if the amount of the added standard (solution of the standard) is negligible as compared with the volume of the liquid
phase.

An analysis of the expression for f_{iLGS} in Eq. (7-110)
reveals that the value of the term $K_{iLG}V_L/K_{iSG}V_S$ will usually
be much larger than that of the term $V_G^{++}/K_{iSG}V_S$. Hence, the
results will be relatively insensitive to variations in the
volume of the gaseous phase in the loop.

It is not difficult to deduce that the double-sampling
and single-sample standard-addition techniques also are readily
applicable in combination with closed-loop equilibration-trapping head-space gas analysis. The corresponding relations are,
respectively,

$$W_i = \frac{W_{iT0}}{1 - (W_{iT1}/W_{iT0})} \tag{7-119}$$

and

$$W_i = \frac{W_s - W_{iT}}{(W'_{iT}/W_{iT}) - 1} \tag{7-120}$$

In Eq. (7-119), W_{iT0} and W_{iT1} are the weight amounts of
component i accumulated in the trapping tube in the first and
second samplings, respectively; the meaning of the symbols in
Eq. (7-120) is the same as in Eq. (7-116). Again, the amounts
W_{iT0}, W_{iT1}, W'_{iT}, and W_{iT} can be determined by virtue of separate calibrations, or, under identical chromatographic condi-
tions, $W_{iT1}/W_{iT0} = A_{iT1}/A_{iT0}$ and $W'_{iT}/W_{iT} = A'_{iT}/A_{iT}$. With both
the double sampling and single-sample standard-addition

techniques, the identical trapping tube must be employed in both
analytical steps, the system being kept at the same temperature
in both cases.

IV. LIQUID EXTRACTION AND HEAD-SPACE GAS ANALYSIS
OF NONFLUIDIC MATERIALS

In the preceding treatment, only liquid materials were
considered as subjects to GC analysis involving liquid or gas
extraction of the material. However, it can easily be inferred
from the concepts presented that the techniques of liquid extraction and head-space gas analysis should also be applicable
to the analysis of diverse solid, quasi-fluidic, and hetero-
geneous materials. For instance, spots extruded from thin-
layer chromatograms, foodstuffs, pharmaceuticals, products of
forensic interest, biological tissues, plants, etc., might be
subjected to quantitative GC analysis involving extraction or
head-space gas sampling.

With some slight formal modifications concerning the system factor and the concentration units to be used for expressing the results, most of the techniques discussed in Sections
II and III of this chapter can readily be applied to the
analysis of nonfluidic materials as well. These modifications
result from the fact that the amount of nonfluidic materials
is specified by weight rather than by volume.

A. *Liquid Extraction*

Let us denote the sample by $\underline{X}$; the weights of the sample
and of component i contained in it will be $W_{(i)}$ and W_i. After
mixing the sample with the extrahent, the amount W_i will divide itself between the extract and the parent material by

$$W_i = W_{i\underline{e}} + W_{i\underline{X}} \qquad\qquad (7\text{-}121)$$

where $W_{i\underline{e}}$ and $W_{i\underline{X}}$ are the weights of component i in the extract

and in the parent material. In this case, the partition coefficient will be defined as

$$K_{i\underline{xe}} = \frac{W_{i\underline{x}}/W_{(i)\underline{p}}}{W_{i\underline{e}}/V_{(i)\underline{e}}} \tag{7-122}$$

where $W_{(i)\underline{p}}$ is the weight of the parent material and $V_{(i)\underline{e}}$ is
the volume of the extract. Hence, Eqs. (7-121) and (7-122)
give

$$W_i = W_{i\underline{e}}\left(\frac{K_{i\underline{xe}}W_{(i)\underline{p}}}{V_{(i)\underline{e}}} + 1\right) = W_{i\underline{e}}\delta_{i\underline{pe}} \tag{7-123}$$

so that

$$\frac{W_i}{W_{(i)}} = g_i = \frac{W_{i\underline{e}}}{W_{(i)}}\delta_{i\underline{pe}} \tag{7-124}$$

where $\delta_{i\underline{pe}}$ again stands for a system factor.

If trace analysis is concerned [$W_i \ll W_{(i)}$] and the major
components of the sample are insoluble in the extrahent, then
the weight $W_{(i)\underline{p}}$ and volume $V_{(i)\underline{e}}$ are practically identical
with the weight of the sample and the volume of the extrahent,
respectively. The considerations presented in Section II of
this chapter could now be resumed in light of Eqs. (7-123) and
(7-124). This would result in the replacement of $f_{i\underline{pe}}$, $f_{s\underline{pe}}$,
$V_{(i)}$, $V_{(i)}^*$, and $V_{(s)}$ by $\delta_{i\underline{pe}}$, $\delta_{s\underline{pe}}$, $W_{(i)}$, $W_{(i)}^*$, and $W_{(s)}$,
respectively, and of q_i by g_i in the equations concerned, thus
adapting them to analytical problems involving extraction of
nonfluidic materials with liquid extrahents. All the condi-
tions discussed regarding the ratio $V_{(i)\underline{p}}/V_{(i)\underline{e}}$ and the system
factor $f_{i\underline{pe}}$ apply in the same manner to the ratio $W_{(i)\underline{p}}/V_{(i)\underline{e}}$
and the system factor $\delta_{i\underline{pe}}$

B. *Head-Space Gas Analysis*

The concepts of analysis of the gas equilibrated in a
closed system with a nonfluidic material as the condensed
phase can be based on considerations quite analogous to those
discussed for liquid extraction of this kind of material.
Hence, we can write

$$W_i = W_{iG} + W_{iX} \tag{7-125}$$

$$K_{iXG} = \frac{W_{iX}/W_X}{W_{iG}/V_G} \tag{7-126}$$

and

$$W_i = W_{iG}\left(\frac{K_{iXG}W_X}{V_G} + 1\right) = W_{iG}\delta_{iXG} \tag{7-127}$$

where W_{iG} and W_{iX} are the amounts of component i in the gas
and in the condensed material, V_G and W_X are the volume of the
gaseous phase and the weight of the condensed phase, and the
meaning of the other symbols is known. The implications of
defining the amount of the sample to be analyzed and of the
standard substance by weight rather than by volume are the
same as in the case of liquid-extraction analysis.

With both liquid-extraction and head-space gas analysis,
it may be impracticable to define the true volume of the extract and of the gaseous phase, respectively, if the volume
of the coexisting nonfluidic phase cannot be measured. However, with the more important variants involving the use of
the standard-addition technique, the absolute values of $V_{(i)e}$
and V_G need not be known; the only requirement is that the
ratios $W_{(i)p}/V_{(i)e}$ and W_X/V_G be the same in the analysis of
both the initial and the standardized systems, which can be

assured by keeping the ratio of sample weight to extrahent or
gas-phase volume constant.   Unfortunately, the double-extraction and double-sampling techniques are inapplicable if the
absolute volume of the extract or of the gaseous phase cannot
be defined [cf. Eqs. (7-66), (7-67), and (7-101)].

The alternatives of head-space gas analysis involving the
complete stripping of the gas from the sample can also be readily applied to the analysis of nonfluidic material, with the
reservation that the sample amount is weighed rather than measured by volume.   Thus, in this case, Eqs. (7-102) and (7-106)
will assume the forms

$$g_i = \frac{W_s}{W_{(i)}} \frac{1}{(A'_{is}/A_i) - (W'_{(i)}/W_{(i)})} \qquad (7\text{-}128)$$

and

$$g_i = \frac{q^+_{iG} V^+_{(i)G}}{W_{(i)}} \qquad (7\text{-}129)$$

respectively.

From Eq. (7-116), it is apparent that the closed-loop
combination of head-space gas analysis with the chromatographic equilibration method, as described for the analysis
of gas-liquid systems, can be used without any alterations
also for analyzing systems containing nonfluidic materials.

Chapter 8

QUANTITATION OF THE CHROMATOGRAM

I. MANUAL TECHNIQUES OF PEAK-AREA DETERMINATION

Methods of manual integration of chromatographic peaks
can be classified into two groups; planimetric methods, and
methods of calculating the peak area from linear parameters
of the peak. While planimetric methods are applicable to the
measurement of the areas of planar configurations of any shape,
methods of the second group are limited to peaks approximating
closely a Gaussian.

Essentially, two manual planimetric methods are employed
in gas chromatography, consisting in the direct measurement of
the peak area with the aid of a planimeter or in the determination of the area indirectly by weighing the peak cut out
either from the chromatogram proper or from its copy. In both

cases, the procedure involves an accurate tracing, with the
stylus of the planimeter or with scissors, of the entire perim-
eter of the peak the area of which is to be determined. Hence,
these methods are readily applicable to the measurement of the
areas of peaks showing severe asymmetry, configurations com-
posed of overlapped peaks, the parts of a composite peak resolved by separation lines, etc., in addition to the measurement of symmetrical and well-resolved peaks. With direct
planimetry, the precision of results can be increased by tracing the area several times and taking an average value. In the
cut-and-weigh method, precision depends on the constancy of the
planar density of the paper. Unfortunately, the determination
of peak area by both variants of manual planimetry is tedious
and of limited precision.

The methods for determining peak area from linear parameters of the peak can be verified by virtue of an analysis of
the Gaussian curve.

The analytical significance of peak area has been illustrated in Chapter 4. The analytical significance of the peak
height arises from the relation between peak height and area.
Provided the contour of the peak is a Gaussian curve, a peak
of unit area (normalized Gaussian curve) can be described by
[116]

$$h = \frac{1}{\sigma\sqrt{2\pi}} \exp\left[-\frac{1}{2}\left(\frac{b_R - b}{\sigma}\right)^2 \right] \qquad (8-1)$$

where b is the distance from the start, h is the instantaneous
recorder pen deflection at this point, b_R is the distance of
the coordinate of maximum deflection (of peak height) from the
start, h_m is the peak height, and σ is the standard deviation.
For the peak height, dh/db = 0 and

$$h_m = \frac{1}{\sigma (2\pi)^{1/2}} \tag{8-2}$$

From the ratio of area to height for a curve described by Eq. (8-1) (A = 1), we obtain

$$A = h_m \sigma (2\pi)^{1/2} \tag{8-3}$$

a relation which is true for any Gaussian curve. It also indicates that, at a given σ, the peak area is proportional to the peak height. For a deflection corresponding to $\pm\sigma$, h_σ, it can be easily derived ($b_R - b = \pm\sigma$) that

$$h_\sigma = h_m e^{-1/2} \tag{8-4}$$

Simple analysis of Eq. (8-1) shows that the points h_σ, $\pm\sigma$ correspond to the points of inflection of the curve ($d^2h/db^2 = 0$) and that the tangents at these inflection points intersect the b axis (baseline) to form two intercepts at $\pm 2\sigma$. The width of a peak at the points of inflection is then 2σ, and the width between the points of intersection of the two tangents and the base line is 4σ. In practice, it is usual and easier to measure the width at half-height, $\Delta b_{1/2}$. From Eq. (8-1) it may be derived ($b_R - b$ for $h_m/2$) that

$$\Delta b_{1/2} = \sigma (8 \ln 2)^{1/2} \tag{8-5}$$

II. METHODS OF PEAK-AREA CALCULATION

Using Eqs. (8-5), (8-2), and (8-3) it is possible to verify the rule for peak-area calculation as the product of the height and width at half-height. If A_1 is the area computed in this manner, then

$$A_1 = \left(\frac{4 \ln 2}{\pi}\right)^{1/2} \tag{8-6}$$

The calculation of A_1 has been done from the parameters of a
normalized Gaussian curve, whose actual area is unity.   The
immediate result is that

$$A = A_1 \left(\frac{\pi}{4 \ln 2} \right)^{1/2}$$
(8-7)

where the factor $(\pi/4 \ln 2)^{1/2}$ has an approximate value of
1.06.

Similarly, the rule can be verified for area determination
as the area of a triangle whose sides are the tangents at the
points of inflection and the respective intercept on the baseline.   If the base of this triangle is twice the curve width
at the points of inflection, then the height of this triangle
must be twice the h coordinate of the points of inflection,
i.e., $2h_\sigma$.   The width at the base has been said to equal 4σ.
If the area of the triangle is A_2, then, using Eq. (8-4),

$$A_2 = 4\sigma h_m \, e^{-1/2}$$
(8-8)

From the ratio of the actual peak area, as given by Eq.
(8-3) to the area of the triangle, we obtain

$$A = \frac{A_2 (2\pi e)^{1/2}}{4}$$
(8-9)

where the factor $(2\pi e)^{1/2}/4$ has a value of about 1.033.

In addition to the methods expressed by Eqs. (8-3), (8-7),
and (8-9), it has been recommended [117, 118] that the product
of peak height and the distance of its apex from the start,
i.e., $h_m b_R$, be used for peak-area determination.   The justifi-
cation for this procedure can be shown in the following way.
For the standard deviation σ^*, measured as the actual spread
of the zone, the plate theory [119] indicates that

$$\sigma^* = \sqrt{HL} \qquad\qquad\qquad (8\text{-}10)$$

where H is the height equivalent to a theoretical plate and L is column length. Under particular conditions, the time Δt_{σ^*} necessary for the zone to travel a distance σ^* is

$$\Delta t_{\sigma^*} = \frac{\sigma^*}{v^*} \qquad\qquad\qquad (8\text{-}11)$$

where v^* is the velocity of the concentration maximum of the band. The quantity σ is given by $\sigma = (db/dt)\Delta t_{\sigma^*}$, whereupon

$$\sigma = \frac{db}{dt}\frac{(HL)^{1/2}}{v^*} \qquad\qquad\qquad (8\text{-}12)$$

Multiplying the numerator and denominator of the right-hand side by L, $\sigma = (db/dt)(L/v^*)(H/L)^{1/2}$, where L/v^* is the retention time t_R. Since the product $(db/dt)t_R$ equals b_R,

$$\sigma = \left(\frac{H}{L}\right)^{1/2} b_R \qquad\qquad\qquad (8\text{-}13)$$

On combining this with Eq. (8-3),

$$A = h_m b_R (H)^{1/2}\left(\frac{2\pi}{L}\right)^{1/2} \qquad\qquad\qquad (8\text{-}14)$$

It must be emphasized that H is not a constant. When using this method of area determination, it is necessary also to make a correction for possible differences in separation efficiency in addition to the correction for different detector sensitivities.

III. INTERPRETATION OF CHROMATOGRAMS BY PEAK HEIGHTS

Peak heights are much more convenient to measure and, sometimes, also more precise than are area measurements. There are cases, however, where calculations based on peak heights are unsuitable or even impossible. The decision

between peak heights and areas should be based on the following considerations.

A. Character of the Chromatographic Record

If the peaks are narrow and tall, then the height is a
more accurate quantity than the area.

B. Detector Properties and the Stability
of the Working Conditions

In the detection with mass-sensitive/destructive detectors,
the peak area is theoretically independent of the rate of introduction of the substance chromatographed into the sensing element, whereas the height is proportional to this rate. The rate
of mass introduction into the sensing element can be altered by
changing the column temperature or the carrier-gas flow rate.
If sufficient stabilization of these two parameters is not feasible, it is inconvenient to use the heights in calculations
where detection is performed with mass-sensitive/destructive
detectors. When using the other types of detector (MN, CN, CD),
the peak area is proportional to the component concentration in
the column effluent and inversely proportional to the flow rate.
Peak height is also proportional to the component concentration,
but only slightly dependent on the flow rate. Hence, it follows
that if temperature stability is inadequate, which is decisive
for the component concentration in the column effluent, then
calculations using the heights are inappropriate. In contrast,
if flow-rate stabilization is not feasible, calculations with
the areas are not suitable with the above types of detector,
and peak heights are preferred. If both the temperature and
flow rate are easy to stabilize, then it is immaterial whether
the heights or areas are used in the calculations. The only
criterion which could be applied here is the time constant of

the detection system. Narrow and high peaks can suffer from
height distortion caused by a large time constant, whereas the
area remains unaffected (cf. [21, 22]).

C. Peak Shape

For asymmetrical peaks, calculations based on the heights
are inappropriate.

D. Working Technique Used

It is necessary to bear in mind that the total amount of
a substance chromatographed present in the chromatographic band
is generally proportional to the peak area and not to the peak
height. Proportionality between the amount of material and the
height of the peak exists only for a given substance, chromatographed under constant conditions, i.e., in such cases where
there is proportionality between the height and area. It follows from (8-14) that

$$\frac{A_i}{A_s} = \frac{h_{mi} b_{Ri} (H_i/H_s)^{1/2}}{h_{ms} b_{Rs}} \tag{8-15}$$

where h_m is the height of the peak, b_R is the distance of peak-
maximum coordinate from the start, and H is the height equivalent to a theoretical plate. It can be inferred from relation
(8-15) that, if correction factors defined in virtue of merely
the relative molar response are to be applied, [cf. Eq. (6-10)],
then:

1. When employing the technique of absolute calibration,
it is possible to calculate with peak heights only if the compound to be determined is used as the standard substance.
This holds both for the variants of direct comparison and of
constructing a calibration curve.

2. With the internal-standard technique, when employing
the variant of direct comparison, it is impossible to calculate
with peak heights. When proceeding with the aid of a calibration curve, it is possible to use peak heights, but such a calibration curve is applicable exclusively for the given system
and operating conditions and cannot be in any way generalized:
the slope of a calibration curve constructed by means of peak
heights involves a specific factor $(b_{Ri}/b_{Rs})(H_i/H_s)^{1/2}$, where
the ratio H_i/H_s (height equivalents to a theoretical plate) is
characteristic of the substances i and s and is dependent on
all the numerous variables which can control the chromatographic
process.

3. When employing the standard-addition technique, calculation with peak heights is feasible in all cases, since peak-
area ratios of the same substance appear in all the relations.

4. With the internal-normalization technique, the use of
peak heights for the calculation is impracticable.

From the chromatogram of a mixture of known composition,
it is of course possible to calculate easily correction factors
which would make it feasible to calculate with peak heights
even in the cases mentioned under 1, 2, and 4. However, these
factors involve, in addition to 1/RMR or M/RMR, also the expression $1/b_R(2\pi H/L)^{1/2}$, which entails the same qualifications as
mentioned in connection with the internal-standard technique in
the variant of a calibration curve.

IV. AUTOMATIC PROCESSING OF CHROMATOGRAMS

Manual quantitation of chromatograms is tedious, time-con-
suming, and relatively imprecise. Therefore, soon after gas
chromatography had become a widely employed analytical method,
means of automating the operation were sought for. Development

in this respect proceeded from mechanical disk integrators to electronic digital integrators and single-purpose computer systems. Nevertheless, the following situation was revealed [120] in the United States as late as in 1966: 67% of chromatograms were evaluated manually, 21% were integrated with a disk integrator, and only about 12% were processed with the aid of an electronic integrator or computer. The situation has changed markedly in the subsequent years; during 1968 and 1969 alone, more than a hundred laboratories in the United States were equipped for computer-aided processing of chromatograms [121].

A. *Automatic Integration of the Detector Response*

Automatic integrators perform a very difficult operation in the quantitative GC analysis, consisting in the measurement of the areas of the individual peaks. A number of integrating devices based on mechanical or electrical principles have been described [122, 123], but only two types of integrator have found a wider usage: the mechanical disk integrator and electronic digital integrators.

1. *Mechanical Disk Integrator*

The disk-and-ball integrator is probably the most employed mechanical integrating machine. The basic parts of this integrator, as indicated by the name, are a disk of radius R, rotating at a constant angular velocity ω, and a small ball of radius ρ, pressed against the disk. The ball can be moved by means of a draw bar to any position r(t) along the radius of the disk from the center to the value R. The ball, rolling on the disk, revolves at a rate given by the angular velocity of the disk, the time-dependent deflection $0 \leq r(t) \leq R$, and the radius ρ. The total number of revolutions of the ball, N, made within the time interval $\langle 0, t \rangle$ is

$$N = \frac{\omega}{2\pi\rho} \int_0^t r(t) \, dt \tag{8-16}$$

and is picked up mechanically by a camshaft and recorded by an
auxiliary pen of the recorder. The operation of this integrator is closely associated with the recorder and, consequently,
reflects all the shortcomings of the latter. Nevertheless, the
integrator works linearly within the entire range of the recorder scale and, when the detector attenuation is controlled properly, affords a precision of 0.9-1.3%. Accurate zeroing of the
detection and recording systems and a stable baseline are basic
prerequisites for proper functioning. The manufacturers provide facilities for the elimination of systematic error due to
baseline drift, but their use increases the time of analysis
and decreases the precision of results. In the most advanced
version of this integrator, the revolutions of the ball are
picked up photoelectrically and the values of N corresponding
to the individual peaks are printed.

2. *Electronic Digital Integrators*

The heart of these integrators is a voltage-to-frequency
(VF) converter, which transforms the detection signal, after
suitable processing by an operational amplifier, into a series
of pulses the frequency of which is proportional to the voltage. The pulses are summed in a counter, usually combined with
a buffer memory, and the sums are fed, via an output converter,
to a printer. Between the converter and the counter is a gate
controlled by the logic of the integrator. A functional diagram of an electronic digital integrator is shown in Fig. 10.
The detector signal is fed first to the operational amplifier,
from which, after some processing, it is fed to the recorder,

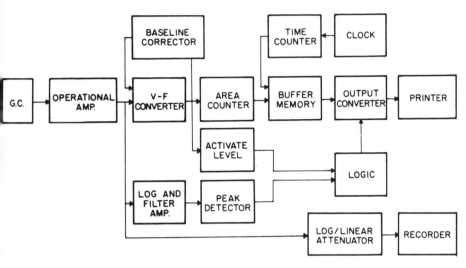

FIG. 10.   Functional diagram of an electronic digital
integrator.

VF converter, and peak detector. The ratio of frequency to input voltage is about 1 kHz/mV. The functioning of the instrument is governed by the control logic, which receives information from both the converter and the peak detector. The logic
controls the automatic baseline corrector and the counters (peak
integrals and retention times). On a command from the logic,
the data from the counters are transferred to the buffer memory,
the counter gates are closed, and the output converter scans the
memory; the output data are printed, punched into tape, or sent,
in an appropriate code, directly to a computer. Retention times
are registered by means of a constant-frequency generator and
are printed either independently or simultaneously with the
peak areas.

The peak detector is usually a differentiating circuit
the output voltage of which is proportional to the instantane-

ous value of the time-differentiation of the signal. The maximum sensitivity of this circuit is usually ± 0.1 $\mu V/sec$ and
can be attenuated appropriately by means of control elements
on the front panel of the instrument. Between the differentiation circuit and the output of the operational amplifier,
there are always inserted filters for the elimination of noise;
the parameters of the filters are also adjustable.

Most integrators of the above kind are provided with an
automatic baseline corrector. The corrector monitors the baseline and compensates for drift by feeding back an equal and
opposite voltage to the input of the VF converter. The maximum
rate of baseline-drift compensation is adjustable and can be
varied within 0.1-10 $\mu V/sec$. At the moment the differentiating
circuit recognizes the beginning of a peak and gives the command for integration, the baseline tracker is blocked and the
compensation voltage remains unchanged during the peak
integration.

The gate of the peak-integral counter is opened by the
logic at the moment when the peak detector has sensed a positive slope. The end of integration occurs as soon as the logic
condition RC X N + L = "0" is fulfilled (RC, N, and L stand for
"running counter," "negative slope," and "level"); RC = "1" if
the counter is running, N = "1" if the differentiation of the
signal is negative, and L = "1" if the signal exceeds an adjustable threshold value. Thus, the gate of the counter will be
closed in two cases: (1) the counter is running (the signal is
above the baseline), but the differentiation of the signal is
no longer negative; (2) the counter has stopped (the signal has
dropped to the level at which peak integration started) and the
differentiation of the signal is or is not negative.

There are integrators with somewhat more sophisticated
logic, affording integration on the tangent of a peak tail.

The tangent is formed automatically if the signal drops during integration below the level existing at the start of integration. In this case, the end of integration is independent of the level of the signal and is determined exclusively by the fall of the negative slope below a preset value.

The precision of integration depends on the correctness of the logic decisions. These decisions, however, are determined by parameters that can be varied within fairly wide limits by the operator. Hence, the operator's experience, along with the quality of the chromatogram, determines the precision of the results. The logic of electronic digital integrators and the effect of the individual adjustable parameters on the precision of integration were analyzed by Karohl [124] and Baumann and Tao [125].

A great advantage of electronic digital integrators is the large linearity range of the analog-to-digital conversion; with good instruments, the deviations from linearity do not exceed ±2 µV within the working input-voltage limits of 0-1 V.

B. *Computer-Aided Processing of Chromatograms*

The automatic integrators discussed in the preceding subsection have certain disadvantages. The capability of their control logic hardware to manage with complicated chromatograms is rather limited. Further, in a given time period, the integrator is able to process input data from only a single chromatograph, and the price of the integrator is relatively high, comparable to that of a good gas chromatograph. The results are merely the raw integrals of the individual peaks and their retention times, rather than an analytical report. It was mainly these disadvantages that led, along with decreasing price of computers, to the use of computers in chromatographic laboratories. A number of variants of computer-aided processing of

chromatograms have been described [126-128]. The individual
variants differ from each other in both the method of data
transfer between the chromatograph and the computer (off-line
and on-line systems) and the kind of transferred data (hybrid
and pure digital systems).

The simplest setup is a combination of chromatograph, integrator, and computer. This is a hybrid system (the logic of
the integrator works by virtue of analog control data); the
task of the computer is limited to searching in the library of
data on standards for the names and response factors of the
components analyzed, stored at the periphery of the computer,
and to processing the corrected peak integrals according to
the chosen and programmed technique of quantitative GC. The
results are delivered in the form of printed analytical report.
Identification of individual components (name and response factor) is carried out by virtue of retention data (retention time,
retention index) and their permissible dispersion (so-called
window). As the transfer of reduced data is slow, a direct
connection of the computer and the integrator, or several integrators, is unsuitable from the economic viewpoint. Usually
an off-line system is employed in which the integrator output
data are punched in real time into tape; batches of tape are
processed intermittently with the computer. An exception to
this rule is presented by computers adapted for work in a time-
sharing mode; such computers can perform other programs between
entries of integrator data. Another exception is a case in
which the computer is replaced by a programmable calculator
adapted for direct connection with the integrator. In this
case, the exception is justified by the relatively low price
of the calculator on the one hand and, on the other hand, by
the limited performance capabilities of the latter. The disadvantages of integrators, however, manifest themselves in the
hybrid systems as well.

A more sophisticated configuration is constituted by pure
digital systems. The analog signal of the detector is transformed via an analog-to-digital converter into a sequence of
digital data compatible with the input system of the computer.
Detection and integration of peaks as well as measurement of
their retention times is performed numerically. The other
operations performed by the computer are analogous to those
mentioned with hybrid systems. In this case, the quality of
the processing of chromatograms is determined by the flexibil-
ity and suitability of the program for processing the input
data; this program replaces the logic of the integrator. In
automatic processing of routine analyses, usually several
chromatographs are connected to a computer by means of a multi-
plexer, and the signals of all the chromatographs are processed
simultaneously in real time. There are available on the market
a number of single-purpose computers equipped with the software
for automatic simultaneous processing of chromatographic data
from several chromatographs.

In more sophisticated cases, input data representing the
time course of the signal are stored in the peripheral memory
of the computer and are processed by the program after the
analysis has been finished. As the data can pass through the
processing program several times, very complicated chromatograms with a number of overlapped peaks can successfully be
processed in this way. A replicate processing of the same
chromatogram can be carried out also in an off-line mode if
the chromatogram is recorded in the form of punched or
magnetic tape.

The highest organization of automatic processing of chromatograms is represented by hierarchical computer systems, consisting in one or more small single-purpose satellite computers
and one large central computer with a large-capacity core memory and a high computing speed. The satellite computers

process data of several chromatographs or spectral analyzers
in real time and enter the intermediate results to the central
computer for final processing.

V.   RESOLUTION OF OVERLAPPED PEAKS

Even with a well deliberated choice of the sorbent and
with the use of high-efficiency columns, the analyst is often
confronted with chromatograms containing overlapping peaks.
It is obvious that complete quantitation of a chromatogram involving overlapped peaks is feasible only if the composite peak
area can be accurately separated into its constituents.   This
may be a rather difficult task.   Several methods have been de-
vised for resolving the areas of fused peaks, based either on
different techniques of linear separation of the composite peak
area or on constructing mathematically a synthetic chromatogram
and fitting it iteratively to the composite peak by nonlinear
regression analysis.   While linear separation of peaks can
easily be applied to manual evaluation of chromatograms, even
though some of them have been widely employed to serve the
basis for commercial computer-based peak-separation programs,
the methods of iterative curve-fitting are applicable only with
the aid of computers.   Some linear peak-separation techniques
are illustrated in Fig. 11.

Both of the approaches we have outlined afford merely better or worse approximations to the true peak areas.   A common
handicap of all the techniques of linear separation is that the
correct allocation of the separation line or lines depends on
the degree of overlap and the relative sizes of the overlapped
peaks.   In other words, the allocation of the separation lines
just as specified by the definition of the given technique will
always deviate more or less from the allocation that would
divide the composite peak area into its true components, the

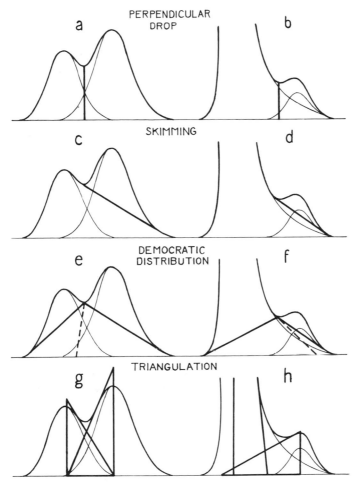

FIG. 11.   Representation of the methods of linear separation of overlapped peaks.

extent of the deviation being generally a function of the peak-
resolution factor and the relative sizes of the overlapping
peaks.   Theoretically, this function could serve as the basis
for expressing correction factors.   A system of such correction
factors, based on a Gaussian peak-shape model, was worked out
for the perpendicular-drop method [129], but it proved to be of
limited utility.   If the area produced by two partially fused
perfect Gaussian peaks of unlike areas is divided into two
parts by a perpendicular dropped from the minimum of the valley
to the baseline [Fig. 11 (a), (b)] as specified by the perpendicular-drop method, the area of the smaller part should always
show a negative deviation as compared with the area of the corresponding original peak, the deviation increasing with the degree of overlap and the ratio of the peak heights.   Hence, with
this idealized model, the results obtained with the perpendicular-drop method should suffer systematically from a negative
error for the smaller peak and a positive error for the larger
one, no matter whether the smaller peak precedes the larger one
or vice versa.   However, with a real chromatogram the situation
is different.   It was found experimentally [130] that, if the
smaller peak precedes the larger one, the bias of the smaller
part of the composite peak area is indeed negative, and application of theoretically predicted correction factors may improve
the accuracy of the results.   However, if the smaller peak is
located after the larger one, the area of the smaller peak, as
determined by the perpendicular-drop method, shows a significant
positive error.   Hence, the use of the correction factors would
even increase the systematic error in this case.   This phenomenon is obviously due to the tailing of the larger peak, viz.,
the true area of a small peak superimposed on the tail of a
large one may be comparable with or even much smaller than the
area of the corresponding segment of the tail.   Such a situation
is apparent from Fig. 11 (b), for instance.   As the degree of

tailing depends on many factors, such as the properties of the
column, temperature of the column oven and of the injection
port, size of the sample charge and the method of its introduction into the gas chromatograph, carrier-gas flow rate, etc.,
it is hardly possible to find any useful regularities that
could provide for the correction of this kind of error.

As the difference in the heights of the incompletely resolved peaks increases at a given degree of overlap, the tail
of the larger peak becomes virtually a descending baseline for
the succeeding small peak, and a good approximation can be obtained by employing the tangent (skimming) method [Fig. 11 (d)].
However, with fused peaks of comparable areas, this method
would give completely false results (cf. [Fig. 11 (c)]). Separation by the skimming method can further be improved by the
so-called democratic distribution of the area under the tangents [131]. In this case, tangents are drawn under both peaks
from the minimum of the valley to both ends of the composite
peak, the area formed by the tangents and the baseline is divided in proportion to the areas above the tangents, and the
portions so obtained are added to the skimmed areas. Figure 11
(e) and (f) show that, while the separation of a small peak from
the tail of a large one is improved by this procedure, the separation of peaks of comparable sizes will be laden with gross
errors.

With the triangulation method [132], the region of the
overlapped parts of the peaks is separated in proportion to the
areas of the right-angle triangles drawn as shown in Fig. 11
(g) and (h). This method may give results similar to those
obtained by dropping a perpendicular, but, likely, with a lower
accuracy and precision.

A linear peak-separation method based on a model in which
both overlapped peaks were replaced by isosceles triangles has
also been proposed [133]. Though this method allows the

position of the separating perpendicular to be calculated from
merely the peak height ratio and the horizontal spacing between
the peak maxima, it can lead to gross errors when applied to
real chromatograms [130].

It can be inferred from Fig. 11 that, if a linear peak-
separation method is to be employed, overlapped peaks of comparable sizes are best separated by the plain perpendicular-
drop method, while peaks of grossly different sizes are better
separated by the skimming method, combined with the democratic-
distribution step if need be.

The methods of iterative curve fitting are more reliable
than those of linear separation, though there are also certain
limitations with the former as to the accuracy and precision of
the results. These limitations stem from the necessity to define an adequate peak-shape model and to estimate the initial
parameters of the peaks to be determined. Procedures involving
the use of a Gaussian peak with different widths of the leading
and the trailing half [134, 135], a Gaussian peak modified with
an exponential decay component [136-138], and experimentally
determined peak shapes [139, 140] have been employed as models
for curve fitting.

Chapter 9

RELIABILITY OF RESULTS

The problem of the reliability of results obtained by gas chromatographic quantitative analysis has been dealt with by many authors. Owing to the large number of variables involved in a quantitative GC experiment, this problem is fairly broad. The degree of reliability of results of quantitative gas chromatography depends on the technique employed and its adaptation to the given analytical task, the quality of the GC instrument, the choice of sorbent, the column, the working conditions, the method of processing the chromatogram, and, particularly, on the experience of the analyst and the care with which he has performed the analysis. In addition, the reliability of results obtained with the given instrumentation operated by the same analyst will obviously depend on the nature of the analytical problems, such as the number, kind, and concentration of

the components to be determined in the sample, the state of
aggregation of the phases constituting the system under analysis, the number of phases, etc.   Hence, it is not surprising
that the findings reported by different investigators on the
reliability of GC quantitation have frequently been at variance.

I.   FUNDAMENTAL STATISTICAL DEFINITIONS

The term reliability refers to the precision and accuracy
of results, the former accounting for the degree of coherency
of the individual results of a set of measurements and the latter for the degree to which the data measured approach the true
value.   An objective criterion of the precision of the results
within a set of n measurements of a quantity X is the estima-
tion of the standard deviation of this quantity, S_X, defined by

$$S_X = \left[\frac{\sum\limits_{i=1}^{n} (X_i - \bar{X})^2}{n - 1} \right]^{1/2} \tag{9-1}$$

where X_i is an individual value from the set of the results and
$\bar{X}$ is the arithmetic mean value of the set.   Precision relates
to the repeatability of the measurements within the set.   The
square of the standard deviation is called variance.   The standard deviation expressed as a fraction of the mean value is
called the relative standard deviation or the coefficient of
variation; denoting the latter by I_X, we have

$$I_X = \frac{S_X}{\bar{X}} \tag{9-2}$$

If the quantity measured is a function of several independent variables, say a, b, c, $\cdots$, k, the standard deviation of
this quantity can be calculated from the variances of the

individual independent variables by virtue of error propagation. Thus, if $X = X(a,b,c, \cdots, k)$, it follows for S_X that

$$S_X = \left[\left(\frac{\partial X}{\partial a}\right)^2 s_a^2 + \left(\frac{\partial X}{\partial b}\right)^2 s_b^2 + \cdots + \left(\frac{\partial X}{\partial k}\right)^2 s_k^2\right]^{1/2} \qquad (9\text{-}3)$$

where S_a, S_b, etc., are the standard deviations of the individual independent variables. For the relative standard deviation, we have

$$I_X = \left[\left(\frac{\partial \ln X}{\partial a}\right)^2 a^2 I_a^2 + \left(\frac{\partial \ln X}{\partial b}\right)^2 b^2 I_b^2 + \cdots\right.$$

$$\left. + \left(\frac{\partial \ln X}{\partial k}\right)^2 k^2 I_k^2\right]^{1/2} \qquad (9\text{-}4)$$

I_a, I_b, etc., being the relative standard deviations of the independent variables.

Accuracy usually refers to the difference between the mean of the set of results and the true value of the quantity measured, $\hat{X}$. Accuracy can be evaluated by testing the statistical significance of the difference $\hat{X} - \bar{X}$, i.e., by comparing the experimental value of the Student criterion, $t_{exptl.}$, with the corresponding tabulated critical value $t_\alpha(n)$ where α is the confidence level. In analytical chemistry, normally a 95% confidence level is accepted, which means that $\alpha = 0.05$. The value of $t_{exptl.}$ is calculated by $t_{exptl.} = n^{1/2}(|\hat{X} - \bar{X}|)/S_X$, and if $t_{exptl.} < t_{crit.}$, the bias is considered as statistically insignificant.

These few statistical definitions have been quoted only to provide a direct reference to the statistical treatment of the techniques of quantitative gas chromatography, presented below. For more detailed information on the problems of statistical processing of analytical results, the reader is referred to the special literature [141-143].

II. SOURCES OF ERRORS

Limitations in the reliability of results stem from various imperfections of sampling, functioning of the instrument, performing the individual steps constituting the technique employed, and processing the chromatogram. Sampling is a basic general problem in analytical chemistry, rendering very diverse situations from case to case as concerns the reliability of analytical results. Owing to relatively small sample charges that are normally introduced into the gas chromatograph, the importance of the potential sources of difficulties in taking a representative sample cannot be over emphasized. These problems have been dealt with in detail in special monographs [144, 145]. As for the reliability of GC instrumentation, modern commercial gas chromatographs, so long as they are operated properly under optimum conditions, do not contribute significantly to the error of the results [146]. Nonetheless, there are cases in which appreciable errors due to the apparatus can be encountered. The instrumental contributions to the errors of GC measurements are discussed in detail elsewhere [8, 19, 21, 22, 147-150].

Perhaps the most thoroughly studied problem concerning the errors in GC quantitative analysis has been that of the evaluation of chromatograms. This problem has also been combined with some studies of the internal-normalization technique [151, 152]. The errors incidental to manual integration of chromatographic peaks have been discussed in detail by Ball et al. [153], who arrived at important conclusions summarized as follows:

1. The error of peak-area measurement depends on the shape of the peak rather than on the method of its measurement; the measurements of peaks of extreme shapes suffer from large relative errors.

2. With any peak, the relative error decreases on
increasing the peak area.

3. With large peaks (50-100 cm^2) having a height of at
least 5 cm, the planimetry method and the height X width at
mid-height method give an error of about 0.5%.

4. For the planimetry method and the height X width at
mid-height method, errors are proportional to $A^{-4/3}$ and $A^{-1/2}$,
respectively.

5. With the height X width at mid-height method, the
error is reduced only slightly on increasing the width above
5 cm, and with planimetry it is of little use to provide peaks
wider than 10 cm.

6. The cutting and weighing method gives dubious results
if the uniformity of the thickness of the paper cannot be guaranteed; otherwise, this method is useful for measuring peaks
of irregular shapes.

7. The triangulation method cannot be recommended under
any circumstances.

8. With irregular peak shapes, methods taking the perime-
ter of the peak into account (direct planimetry, cutting and
weighing) should be employed.

9. So long as the experimental conditions are controlled
precisely, the measurement of peak height alone is most precise.

Later on, the individual methods of manual integration
were reinvestigated and the results compared with those obtained
with the aid of the disk integrator and electronic digital integrators [154]. The precision and accuracy of peak-area determination with electronic digital integrators and in computer-aided
processing of chromatograms have been dealt with by several
authors. Emery [155] tested the inherent precision of an electronic digital integrator by processing repeatedly an electrically simulated chromatogram produced from a high-precision

source of square pulse signals of defined voltage and duration.
This approach made it possible to virtually separate the variability due to the integrator from that due to the chromatographic
system. So long as the duration of the pulses was 10 sec or
more, the relative 2S of the peak area determination in this way
was less than 0.1%. When processing real chromatograms, with
the integrator connected to the gas chromatograph, the precision
was about an order lower. Bauman and Tao [125] evaluated the
errors associated with peak detection, automatic baseline correction, and filtering. They showed the accuracy of peak-area
determination to be related to the quantity SW/H for a Gaussian
peak, where S, W, and H are the slope sensitivity of the peak
detector, peak width, and peak height, respectively. The separate theoretical contributions to the negative systematic error
due to slope detection (with baseline correction switched off)
and due to baseline correction (with the rate of the latter being numerically equal to or higher than slope sensitivity) at
different values of the parameter SW/H are shown in Table 3.
The actual errors measured experimentally were somewhat larger
than those presented in the table. The authors concluded that
the ratio of slope sensitivity to baseline correction rate

TABLE 3

Errors Due to Slope Detection and Baseline Correction
at Different Values of SW/H[a]

$(SW/H) \times 10^3$	Slope detection	Baseline correction
1.0	0.01%	0.35%
10	0.02	0.35
100	0.27	3.4
1,000	10	37

[a]S = slope sensitivity, W = peak width, H = peak height.

should be kept greater than 10. The effect of filtering on the error of peak-area determination was found to be relatively insignificant.

The accuracy and precision attainable in computer-aided quantitation of chromatograms also have been discussed in several papers, aimed particularly at separation of the errors due to the gas chromatograph and due to the computer [156, 157] and at errors incidental to the deconvolution of overlapped peaks [134-137, 158].

A study by Gill et al. [152] showed the following order for the precision of peak-area determination, as expressed in percentage relative standard deviation: (1) planimetry, 4.0%; (2) triangulation, 4.0%; (3) height × width at mid-height, 2.5%; (4) cutting and weighing, 1.7%; (5) disk integrator, 1.3%; (6) electronic digital integrator, 0.4%.

III. PRECISION OF RESULTS OBTAINED BY CONVENTIONAL TECHNIQUES

The mathematical formulas of the individual GC techniques, as presented in Chapter 6, provide for readily applying the concept of error propagation to the expression of relations for calculating the theoretical precision of analytical results from data on the precision of the independent variables [cf. Eqs. (9-3) and (9-4)]. The predictions so obtained can be checked experimentally by processing statistically the results of a set of replicate analyses of model mixtures, carried out by the individual techniques [146].

In some techniques, the ratio RMR_{sr}/RMR_{ir} occurs. This ratio is obviously identical with the quantity RMR_{si} given by $RMR_{si} = (A_s/m_s)/(A_i/m_i)$, A_s and A_i being the peak areas obtained by chromatographing a model mixture of component s (standard) and component i of known molarities m_s and m_i.

Provided that the ratio m_i/m_s has been determined precisely by weighing, we can write

$$s^2_{RMR_{si}} = RMR^2_{si} \left[\left(\frac{S_{A_s}}{A_s} \right)^2 + \left(\frac{S_{A_i}}{A_i} \right)^2 \right] \qquad (9-5)$$

and

$$I_{RMR_{si}} = \left[\left(\frac{S_{A_s}}{A_s} \right)^2 + \left(\frac{S_{A_i}}{A_i} \right)^2 \right]^{1/2} \qquad (9-6)$$

All the symbols referring to the techniques discussed hereinafter have exactly the same meaning as in Chapter 6. The entire treatment is performed in terms of molarities or, if need be, mole fractions. Instead of the absolute mole number of the standard substance, N_s, occurring in the relations for the internal-standard and standard-addition techniques, the corresponding product $V_{(s)}m_s$ will be considered in this treatment. For brevity, h will be employed instead of h_m to denote the peak height.

A.   *Absolute-Calibration Technique*

Alternative I.   Direct comparison of the peak heights of substance i in the chromatograms of the analyzed and calibration samples.   (Pure substance i is used as the standard.)

This alternative is characterized by the relation

$$m_i = \frac{v_{(s)} h_i}{v_{(i)} h_s} m_s \qquad (9-7)$$

Provided that $v_{(i)} \doteq v_{(s)}$ and $h_i \doteq h_s$ (this situation is assumed in all the cases in which it is relevant), the relative standard deviation of the determination of molarity, I_m, can be expressed by

$$I_m = \left[2(I_v^2 + I_h^2) \right]^{1/2} \tag{9-8}$$

where I_v and I_h are the relative standard deviations of the measurements of the sample volume charged into the gas chromatograph and of the peak height, respectively.

Alternative II.   Direct comparison of the peak areas of substance i in the chromatograms of the analyzed and calibration samples.   (Pure substance i is used as the standard.)

In this case,

$$m_i = \frac{v_{(s)} A_i}{v_{(i)} A_s} m_s \tag{9-9}$$

which leads to

$$I_m = \left[2(I_v^2 + I_A^2) \right]^{1/2} \tag{9-10}$$

where I_A is the relative standard deviation of the peak area.

Alternative III.   Direct comparison of the peak area of substance i in the chromatogram of the sample analyzed with the peak area of a standard substance in the chromatogram of the calibration sample.   (The substance to be determined and the standard are different compounds.)

In this case, it is necessary to calculate with corrected peak areas, i.e., Eq. (6-11) applies.   Provided that also the empirical determination of the respective RMR values is involved in the performance of the technique, the RMR's represent normal variables manifesting themselves in the resultant error as given by Eq. (9-6).   It follows from analogy with the preceding case that

$$I_m = \left[2(I_v^2 + I_A^2 + I_{RMR}^2) \right]^{1/2} \tag{9-11}$$

Alternative IV.   Calibration curve, calculation by peak heights.

The work with a calibration curve is based on the relation

$$m_i = \frac{\bar{K}_h h_i}{v_{(i)}} \tag{9-12}$$

where $\bar{K}_h$ is an empirical constant determined by analyzing a series of samples of defined molarities of the substance under determination.   The results of n such analyses can be processed by means of the relation

$$\bar{K}_h = \frac{1}{n} \sum^n \frac{m_i v_{(i)}}{h_i} \tag{9-13}$$

where m_i represents precise values determined by weighing.   The variance and relative standard deviation of the constant $\bar{K}_h$ are given by

$$S_{\bar{K}_h}^2 = \frac{1}{n}\left(\frac{m_i^2 S_v^2}{h_i^2} + \frac{m_i v_{(i)}^2 S_h^2}{h^4} \right) \tag{9-14}$$

and

$$I_{\bar{K}_h} = \left[\frac{1}{n}\left(I_v^2 + I_h^2 \right) \right]^{1/2} \tag{9-15}$$

With respect to the relation for m_i, we can write

$$S_m^2 = \frac{h_i^2}{v_{(i)}^2} S_{\bar{K}_h}^2 + \frac{\bar{K}_h^2}{v_{(i)}^2} S_h^2 + \frac{\bar{K}_h^2 h_i^2}{v_{(i)}^4} S_v^2 \tag{9-16}$$

and the corresponding relative standard deviation is given by

$$I_m = \left[\frac{1}{n}\left(I_v^2 + I_h^2 \right) + I_v^2 + I_h^2 \right]^{1/2} \tag{9-17}$$

This relation applies to the cases in which a new calibration

curve is provided for each individual determination. If a single calibration curve is used for several analyses, the relation
$I_m = (I_h^2 + I_v^2)^{1/2}$ holds, and the relative standard deviation of
the slope of the calibration curve, $I_{\bar{K}_h}$, will manifest itself as
a fraction of the systematic error.

Alternative V. Calibration curve, calculation by peak
areas; calibration carried out with pure substance i.

The relationship between this alternative and the preceding
one is similar to that between the corresponding variants of
direct comparison. We can immediately write $m_i = \bar{K}_A A_i / v_{(i)}$,
where $\bar{K}_A = (1/n) \sum\limits^{n} [m_s v_{(s)} / A_s]$, so that

$$I_m = \left[\frac{1}{n}\left(I_v^2 + I_A^2 \right) + I_v^2 + I_A^2 \right]^{1/2} \qquad (9\text{-}18)$$

When employing a single calibration curve for a set of analyses,
the term $(1/n)(I_v^2 + I_A^2)$ will be omitted.

Alternative VI. Calibration curve, calculation by peak
areas; calibration carried out with a substance different from
substance i.

In this case, it is again necessary to calculate with corrected peak areas; the procedure can be based on the relation

$$m_i = \bar{K}_A \frac{A_i}{v_{(i)}} \frac{RMR_{sr}}{RMR_{ir}} \qquad (9\text{-}19)$$

where $\bar{K}_A$ is given by the same expression as in the preceding
case. If separate values of RMR_{sr} and RMR_{ir} are used, as determined specially for each individual readout, the relative
standard deviation of m is given by

$$I_m = \left[\frac{1}{n}\left(I_v^2 + I_A^2 \right) + I_v^2 + I_A^2 + 2 I_{RMR}^2 \right]^{1/2} \qquad (9\text{-}20)$$

In the case where the RMR_{si} values are readily available, we can write

$$I_m = \left[\frac{1}{n}\left(I_V^2 + I_A^2\right) + I_V^2 + I_A^2 + I_{RMR}^2\right]^{1/2} \tag{9-21}$$

When working with a single calibration curve, the term $(1/n)$ $(I_V^2 + I_A^2)$ in Eqs. (9-20) and (9-21) will be omitted.

B. *Internal-Standard Technique*

Alternate I. Direct comparison of the peak areas of component i and of the standard.

This technique is described by Eq. (6-19), where $N_s = V_{(s)}m_s$, and the corresponding I_m is given by

$$I_m = \left[2\left(I_V^2 + I_A^2 + I_{RMR}^2\right)\right]^{1/2} \tag{9-22}$$

where I_V is the relative standard deviation of the volumes $V_{(s)}$ and $V_{(i)}$ measured out in the preparation of the sample for analysis. If directly determined RMR_{si} values are available, the term I_{RMR}^2 is not multiplied by 2.

Alternative II. Calibration curve, calculation by peak heights.

This variant is based on the relation

$$m_i = \bar{K}_h' \frac{V_{(s)}}{V_{(i)}} \frac{h_i'}{h_s'} m_s \tag{9-23}$$

where h_i' and h_s' are the peak heights of components i and s in the sample-standard mixture. The empirical constant $\bar{K}_h'$ is obviously given by

$$\bar{K}_h' = \frac{1}{n} \sum^n \left(\frac{m_i}{m_s} \frac{V_{(i)}}{V_{(s)}} \frac{h_s'}{h_i'}\right) \tag{9-24}$$

From the relations for $\bar{K}'_h$ and m_i, it follows that

$$I_{\bar{K}_h} = \left[\frac{2}{n}\left(I_V^2 + I_h^2\right)\right]^{1/2} \tag{9-25}$$

and

$$I_m = \left[\frac{2}{n}\left(I_V^2 + I_h^2\right) + I_V^2 + I_h^2\right]^{1/2} \tag{9-26}$$

Alternative III. Calibration curve, calculation by peak areas; calibration carried out with the same pair of substances i and s.

In this case,

$$m_i = \bar{K}'_A \frac{V_{(s)}}{V_{(i)}} \frac{A'_i}{A'_s} m_s \tag{9-27}$$

where the constant $\bar{K}'_A$, representing actually the ratio RMS_{sr}/RMR_{ir}, is given by

$$\bar{K}'_A = \frac{1}{n} \sum^n \left(\frac{m_i}{m_s} \frac{V_{(i)}}{V_{(s)}} \frac{A'_s}{A'_i}\right). \tag{9-28}$$

These relations lead to

$$I_m = \left[\frac{2}{n}\left(I_V^2 + I_A^2\right) + I_V^2 + I_A^2\right]^{1/2} \tag{9-29}$$

Alternative IV. Calibration curve, calculation by peak areas; calibration and analysis carried out with different standards.

Let us denote the standards used in the calibration and in the analytical runs by s1 and s2, respectively. As the general relation quoted in connection with the method of direct comparison holds true independently of whether s is replaced by s1 or s2, we can write for the calibration curve

$$m_i = \bar{K}'_A \frac{V_{(s1)}}{V_{(i)}} \frac{A'_i}{A'_{s1}} m_{s1} \tag{9-30}$$

and assume that the readout is carried out for values given
either by

$$\frac{A'_i}{A'_{s2}} \frac{V_{(s2)}}{V_{(i)}} \frac{RMR_{s2r}}{RMR_{s1r}} m_{s2} \tag{9-31}$$

or by

$$\frac{A'_i}{A'_{s2}} \frac{V_{(s2)}}{V_{(i)}} RMR_{s2s1} m_{s2} \tag{9-32}$$

The relative standard deviation can then be expressed by either

$$I_m = \left[\frac{2}{n}\left(I_V^2 + I_A^2\right) + 2\left(I_V^2 + I_A^2 + I_{RMR}^2\right) \right]^{1/2} \tag{9-33}$$

or

$$I_m = \left[\frac{2}{n}\left(I_V^2 + I_A^2\right) + 2\left(I_V^2 + I_A^2\right) + I_{RMR}^2 \right]^{1/2} \tag{9-34}$$

respectively. In these relations, the term multiplied by the
factor 2/n will be deleted if only a single calibration curve
is used for a series of determinations.

C. *Standard-Addition Technique*

Alternative I. Direct measurement of the charges of the
original sample and of the sample enriched by a defined addition of the substance to be determined; calculation by peak
heights.

This alternative is represented by

$$m_i = \frac{V_{(s)}}{V_{(i)}} \frac{m_s}{\dfrac{h'_{is}}{h_i} \dfrac{V_{(i)}}{v'_{(i)}}\left(1 + \dfrac{V_{(s)}}{V_{(i)}}\right) - 1} \tag{9-35}$$

where h'_{is} and h_i are, respectively, the peak heights of the component determined in the chromatograms of the enriched and the original sample, $v'_{(i)}$ and $v_{(i)}$ being the corresponding volumes charged into the gas chromatograph. When introducing the designations $h'_{is}/h_i = \eta$, $v_{(i)}/v'_{(i)} = \varphi$, and $V_{(s)}/V_{(i)} = \phi$, the relation for I_m can be expressed in the form

$$I_m = \frac{\eta\varphi}{\eta\varphi(1+\phi)-1} \left[(1+\phi)^2 I_\eta^2 + (1+\phi)^2 I_\varphi^2 \right.$$

$$\left. + \left(\frac{\eta\varphi-1}{\phi\varphi}\right)^2 I_\phi^2 \right]^{1/2} \tag{9-36}$$

Provided that $v_{(i)} \doteq v'_{(i)}$ and $V_{(s)} \doteq V_{(i)}$, $\eta = [1 + (m_s/m_i)]/2$, and I_m is given by

$$I_m = \frac{m_i + m_s}{m_s} \left[I_\eta^2 + I_\varphi^2 + \frac{1}{4}\left(\frac{m_s - m_i}{m_s + m_i}\right) I_\phi^2 \right]^{1/2} \tag{9-37}$$

If, in addition, $m_i \doteq m_s$, we obtain

$$I_m = 2\left[2\left(I_h^2 + I_v^2\right)\right]^{1/2} \tag{9-38}$$

Alternative II. Direct measurement of the charges, calculation by peak areas.

In this case, the ratio h'_{is}/h_i in Eq. (9-35) is replaced by A'_{is}/A_i, so that it is possible under the same presuppositions as in the preceding case to write

$$I_m = 2\left[2\left(I_A^2 + I_v^2\right)\right]^{1/2} \tag{9-39}$$

Alternative III. Comparison with an auxiliary reference substance, calculation by peak heights.

In this alternative, the size of the peak of an auxiliary reference substance (p) serves as a measure of the sample amount injected, and we can write

$$m_i = \frac{V_{(s)}}{V_{(i)}} \frac{m_s}{(h'_{is}/h_i)(h_p/h'_p) - 1} \tag{9-40}$$

where h_p and h'_p are the peak heights of the auxiliary substance in the chromatograms of the original sample and of the sample enriched with a defined amount of the standard (substance i). Following a procedure similar to that used in the first variant of this method and supposing that $v_{(i)} \doteq v'_{(i)}$, $V_{(s)} \doteq V_{(i)}$, and, further, $h_i \doteq h_p$, we arrive at

$$I_m = \left\{ 2I_V^2 + I_h^2 \left[10 \left(\frac{m_i}{m_s}\right)^2 + 12 \left(\frac{m_i}{m_s}\right) + 6 \right] \right\}^{1/2} \tag{9-41}$$

If, in addition, $m_i \doteq m_s$, we obtain

$$I_m = \left(2I_V^2 + 28I_h^2 \right)^{1/2} \tag{9-42}$$

Alternative IV. Comparison with an auxiliary reference substance, calculation by peak areas.

This alternative is defined by Eq. (6-35); under the presuppositions applied in the preceding case, we obtain

$$I_m = \left(2I_V^2 + 28I_A^2 \right)^{1/2} \tag{9-43}$$

D. *Internal-Normalization Technique*

There exists practically only one variant to this technique, which can be defined, for calculating mole fractions (x), by the relation

$$x_i = \frac{A_i/RMR_{ir}}{(A_i/RMR_{ir}) + \Sigma \, (A_j/RMR_{jr})} \tag{9-44}$$

where the summation includes all the components of the mixture
except component i. The respective relative standard deviation
is given by

$$
I_{x_i} = \left(\frac{A_i}{RMR_{ir}} + \sum \frac{A_j}{RMR_{jr}} \right)^{-1} \left[\left(\sum \frac{A_j}{RMR_{jr}} \right)^2 \left(I_{A_i}^2 + I_{RMR_{ir}}^2 \right) \right.
$$

$$
\left. + \sum \frac{A_j^2}{RMR_{jr}^2} \left(I_{A_j}^2 + I_{RMR_{jr}}^2 \right) \right]^{1/2} \tag{9-45}
$$

A typical feature of this technique is the necessity to
evaluate all the components of the mixture analyzed, so that
both the precision and accuracy of the individual determinations are interdependent. This situation leads to a number of
possible alternatives. In order to make the task unambiguous,
it is necessary to introduce further presuppositions. We shall
consider two typical alternatives in this treatment:

Alternative I. The concentration of the component under
determination is considerably higher than the concentrations
of the other components, the total number of components (k) is
small, the concentrations of the minor components are mutually
comparable, and the RMR values of the individual components do
not differ appreciably from each other.

Under these circumstances, the peak areas of the minor
components as well as the respective RMR values will be determined with approximately equal relative error. Further, as
$A_i/RMR_{ir} \gg \sum (A_j/RMR_{jr})$ and $\sum (A_j/RMR_{jr}) \doteq (k - 1)(A_j/RMR_{jr})$,
it is possible to write

$$
I_{x_i} = \frac{A_j}{A_i} \left[(k - 1)^2 (I_{A_i}^2 + I_{RMR_{ir}}^2) + (k - 1)(I_{A_j}^2 + I_{RMR_{jr}}^2) \right]^{1/2}
$$

$$
\tag{9-46}
$$

It also follows from the above presuppositions that $I_{A_j}^2 \gg I_{A_i}^2$ and $I_{RMR_{ir}}^2 \doteq I_{RMR_{jr}}^2 = I_{RMR}^2$, so that we have for a binary mixture ($k = 2$)

$$I_{x_i} = \frac{A_j}{A_i} (I_{A_j}^2 + 2I_{RMR}^2)^{1/2} \tag{9-47}$$

In case of a greater number of minor components, while keeping the other presuppositions unchanged, we obtain

$$I_{x_i} = (k - 1) \frac{A_j}{A_i} (I_{A_i}^2 + I_{RMR}^2)^{1/2} \tag{9-48}$$

The consequences resulting from $I_{A_j}^2$ being much greater than $I_{A_i}^2$ are obviously outweighed by the fact that $(k - 1)^2 I_{A_i}^2 \gg (k - 1) I_{A_j}^2$ and $(k - 1)^2 I_{RMR_{ir}}^2 \gg (k - 1) I_{RMR_{jr}}^2$. Thus, under the circumstances quoted above

$$x_i = \frac{A_i / RMR_{ir}}{(A_i / RMR_{ir}) + (k - 1)(A_j / RMR_{jr})} \tag{9-49}$$

and I_{x_i} can be expressed in the form

$$I_{x_i} = \frac{1 - x_i}{x_i} (I_{A_i}^2 + I_{RMR}^2)^{1/2} \tag{9-50}$$

It follows from these relations that the significance of the relative error of the peak areas of minor components decreases to such an extent on increasing the number of components that the error of the peak area of the major component prevails.

Alternative II. The component determined represents a small fraction of the mixture analyzed, and the presuppositions introduced with the first alternative are again applicable.

In this case, again $A_i / RMR_{ir} \gg \Sigma (A_j / RMR_{jr})$, and if the components of the main part of the sample are present in mutually comparable concentrations, we can write $\Sigma \ (A_j/RMR_{jr})$ $\doteq (k - 1)(A_j/RMR_{jr})$. In this case, the general relationship for I_{x_i} will acquire the form

$$I_{x_i} = \left[I^2_{A_i} + I^2_{RMR_{ir}} + \frac{1}{k - 1}\left(I^2_{A_j} + I^2_{RMR_{jr}} \right) \right]^{1/2} \qquad (9\text{-}51)$$

For a small k, such as in the limiting case of a binary mixture, we obtain

$$I_{x_i} = \left(I^2_{A_i} + 2I^2_{RMR} \right)^{1/2} \qquad (9\text{-}52)$$

as $I^2_{A_j} << I^2_{A_i}$. For a large k,

$$I_{x_i} = \left(I^2_{A_i} + I^2_{RMR} \right)^{1/2} \qquad (9\text{-}53)$$

IV. PRECISION OF THE QUANTITIES
REPRESENTING INDEPENDENT VARIABLES

The expressions for the individual variants of the calculation of molarity by conventional techniques of quantitative gas chromatography involve the following independent variables: $v_{(i)}$, $v_{(s)}$, $V_{(i)}$, $V_{(s)}$, h_i, h_s, A_i, A_s, RMR_{ir}, RMR_{is}, and m_s. In order to calculate the actual precision of the results obtained by these techniques, it is necessary to know the actual values of I_v, I_V, I_h, I_A, and I_{RMR}. The value of I_{m_s} is supposed to be negligible as compared to the other I values, as the quantity m_s is determined with a relatively high precision.

The basic criterion of precision is the absolute standard deviation. In a detailed analysis of the techniques of quantitative gas chromatography [146], the values of S_v, S_V, S_h, S_A, and S_{RMR} were determined by replicate measurements of the respective quantities under conditions normally encountered in quantitative GC analyses. The S_v was determined by weighing

doses of tetrabromoethane measured out with a Hamilton 701-N
microsyringe and injected into a weighing bottle adapted as a GC
inlet port. The S_V was determined by weighing doses of toluene
(the latter was employed as a solvent in further experiments)
measured out with a 5-ml pipette. As the study was designed for
calculation with peak heights and the peak areas determined as
the product of the peak height and the peak width at half-height,
the standard deviation of length, S_ℓ, was determined by measuring length etalons with a rule used for the measurement of chromatographic peaks. The standard deviation of the deflection of
the recorder pen, S_r, was accepted to be 3×10^{-2} cm according
to the manufacturer's specifications. Hence, S_h was calculated
by $S_h = (S_\ell^2 + S_r^2)^{1/2}$, and the standard deviation of the peak
width at half-height, S_b, was supposed to equal S_ℓ. Thus, the
S_A was calculated by $S_A = (h^2 S_b^2 + b^2 S_h^2)^{1/2}$, b being the peak
width as specified above. The I_A was then given by $I_A = (I_h^2 + I_b^2)^{1/2}$. The values of S_{RMR} and I_{RMR} were determined by virtue
of Eqs. (9-1) and (9-2).

As the relative standard deviation is referred to the actual
magnitude of the quantity measured, certain rated values of the
variables tested have to be specified in order to make the situation unequivocal. Such values, characterizing unfavorable,
medium, and favorable conditions, are quoted along with the
respective I values in Table 4. Table 5 shows a comparison of
the I_m values calculated from the I's of the independent variables (theor.) with those determined directly from replicate
analyses of model mixtures, performed under the conditions stated
(exptl.). The analyses were carried out on a commercial analytical gas chromatograph with flame-ionization detection under iso-
thermal conditions. The fairly good agreement between the
theoretical and experimental I_m values, especially as obtained
for the favorable and medium conditions, reveals that the contribution of the GC instrument to the resultant error was
insignificant.

TABLE 4

Absolute and Relative Standard Deviations of Basic
Independent Variables in Quantitative Gas Chromatography

Variable	Standard deviation	Unfavorable conditions		Medium conditions		Favorable conditions	
		Rated value	I (%)	Rated value	I (%)	Rated value	I (%)
Volume injected (μl)	4.0×10^{-2}	1	4.0	5	0.8	10	0.40
Volume mixed (ml)	6.0×10^{-3}	0.2	3.0	2	0.3	5	0.12
Peak height (cm)	$3.6 \times 10{-2}$	2	1.8	12	0.3	18	0.20
Peak half-width (cm)	2.0×10^{-2}	0.5	2.5	4	0.5	10	0.20
Peak area (cm^2)	--	1	3.1	48	0.58	180	0.28
RMR_{si}	--	--	4.3	--	0.83	--	0.40

TABLE 5

Precision of Results Obtained by Quantitative Gas Chromatography

Technique	Alternative			Value of I_m					
				Unfavorable conditions		Medium conditions		Favorable conditions	
				Theor.	Exptl.	Theor.	Exptl.	Theor.	Exptl.
Absolute calibration	Direct comparison	s = i	h	5.8	4.4	1.2	1.4	0.65	0.74
			A	6.7	6.3	1.4	1.4	0.70	0.90
		s ≠ i	A	7.7	7.0	1.6	2.0	0.80	0.99
	Calibration curve	s = i	h	4.3	4.3	0.83	0.57	0.50	0.40
			A	5.1	4.9	0.97	0.92	0.54	0.50
		s ≠ i	A	6.4	5.0	1.3	1.2	0.66	0.57
Internal standard	Direct comparison	(s ≠ i)	A	5.9	4.8	1.1	1.2	0.57	0.50
	Calibration curve	s1 = s2	h	3.7	3.4	0.46	0.55	0.32	0.30
			A	4.4	4.3	0.70	0.71	0.39	0.38
		s1 ≠ s2	A	6.9	6.0	1.1	1.1	0.54	0.53
Standard addition	Measuring of charges	(s = i)	h	11	9.6	2.4	2.6	1.3	1.4
			A	13	10	2.7	2.8	1.4	1.4
	Auxiliary substance	(s = i)	h	6.8	5.6	1.6	1.9	1.1	1.4
			A	15	10	3.0	2.6	1.7	2.0
Internal normalization	i = Major component		A	1.2	0.53	0.26	0.32	0.12	0.11
	i = Minor component		A	4.8	2.7	0.92	1.2	0.48	0.58

REFERENCES

1. A. T. James and A. J. P. Martin, Biochem. J., 50, 679 (1952).

2. J. Janák, Chem. Listy, 47, 817 (1953).

3. S. C. Bevan and S. Thornburn, Chem. Br., 1, 206 (1965).

4. A. B. Littlewood, Gas Chromatography, Academic Press, New York, 1962, p. 161.

5. N. L. Gregory and J. E. Lovelock, Anal. Chem., 33, 45A (1961).

6. L. Giuffrida, J. Assoc. Offic. Agr. Chem., 47, 1112 (1964).

7. R. L. Martin and J. A. Grant, Anal. Chem., 37, 644 (1965).

8. R. Kaiser, Chromatographie in der Gasphase, vol. 4, Bibliographisches Institut, Mannheim, 1964, p. 11.

9. H. H. Hausdorff, in Vapour Phase Chromatography (D. H. Desty, ed.), Butterworths, London, 1957, p. 377.

10. J. Novák, Thesis, Technical Univ., Prague, Czechoslovakia, 1965, p. 83.

11. C. H. Heft, in Gas Chromatographie 1958 (H.-P. Angele, ed.), Akad. Verlag, Berlin, 1958, p. 299.

12. E. M. Fredericks and F. R. Brooks, Anal. Chem., 28, 297 (1956).

13. R. L. Grob, D. Mercer, T. Gribben, and J. Wells, J. Chromatogr., 3, 545 (1960).

14. L. C. Browning and J. O. Watts, Anal. Chem., 29, 24 (1957).

15. L. J. Nunez, W. H. Armstrong, and H. W. Cogswell, Anal. Chem., 29, 1164 (1957).

16. S. Dal Nogare and L. W. Safranski, Anal. Chem., 30, 894 (1958).

17. G. Schomburg, Z. Anal. Chem., 164, 147 (1958).

18. M. Rosie and R. L. Grob, Anal. Chem., 29, 1263 (1957).

19. R. S. Evans and R. P. W. Scott, Chimie, 17, 137 (1963).

20. J. C. Sternberg, Advan. Chromatogr., 2, 205 (1966).

21. K. Hána, Coll. Czech. Chem. Commun., 32, 968, 981 (1967).

22. I. G. McWilliam and H. C. Bolton, Anal. Chem., 41, 1755, 1762 (1969); 43, 883 (1971).

23. J. C. Sternberg, in Gas Chromatogr. Symp., 4th, 1963 (L. Fowler, ed.), Academic Press, New York, 1963, p. 161.

24. A. L. LeRosen, J. Amer. Chem. Soc., 67, 1683 (1945).

25. A. J. P. Martin and R. L. M. Synge, Biochem. J., 35, 1358 (1941).

26. J. C. Giddings, Dynamics of Chromatography, Marcel Dekker, New York, 1965, p. 271.

27. F. Čůta, Přednášky o fysikálních a speciálních metodách analytických, Technical Univ., Prague, 1963, p. 6.

28. S. Dal Nogare and R. S. Juvet, Gas-Liquid Chromatography, Wiley (Interscience), New York, 1962, p. 188.

29. I. Halász, Anal. Chem., 36, 1428 (1964).

30. R. P. W. Scott, Nature (London), 176, 793 (1955).

31. H. D. Condon, P. R. Scholly, and W. Averill, Intern. Symp. Gas Chromatogr., Edinburgh, 1960, preprints, p. E1.

32. J. E. Lovelock, Intern. Symp. Gas Chromatogr., Edinburgh, 1960, preprints, p. 9.

33. J. Novák, J. Gelbičová-Růžičková, S. Wičar, and J. Janák, Anal. Chem., 43, 1996 (1971).

34. G. A. Shakespear, Proc. Phys. Soc. (London), 33, 163 (1921).

35. E. G. Hoffman, Z. Anal. Chem., 164, 182 (1958).

36. L. J. Schmauch and R. A. Dinnerstein, Anal. Chem., 32, 343 (1960).

37. E. G. Hoffman, Anal. Chem., 34, 1216 (1962).

38. H. Luy, Z. Anal. Chem., 194, 241 (1963).

39. B. D. Smith and W. W. Bowden, Anal. Chem., 36, 82 (1964).

40. J. Novák, Thesis, Technical Univ., Prague, Czechoslovakia, 1965, p. 30.

41. A. Wassiljewa, Phys. Z., 5, 737 (1904).

42. F. Van De Craats, in Gas Chromatography (D. H. Desty, ed.), Butterworths, London, 1958, p. 248.

43. A. E. Messner, M. Rosie, and P. A. Argabright, Anal. Chem., 31, 230 (1959).

44. N. Hara, S. Yamano, Y. Kumagaya, K. Ikebe, and K. Nakayama, Kogyo Kagaku Zasshi, 66, 1801 (1963).

45. S. Claesson, Arkiv Kemi, Min. Geol., 23A, 1 (1946).

46. A. J. P. Martin and A. T. James, Biochem. J., 63, 138 (1956).

47. C. W. Munday and G. R. Primavesi, in Vapour Phase Chromatography (D. H. Desty, ed.), Butterworths, London, 1957, p. 146.

48. A. G. Nerheim, Anal. Chem., 35, 1640 (1963).

49. A. Liberti, L. Conti, and V. Crescenzi, Nature (London), 178, 1067 (1956).

50. J. I. Henderson and J. H. Knox, J. Chem. Soc., p. 2299 (1956).

51. P. Bullock, in Gas Chromatography (D. H. Desty, ed.), Butterworths, London, 1958, p. 175.

52. G. R. Primavesi, in Gas Chromatography (D. H. Desty, ed.), Butterworths, London, 1958, p. 176.

53. I. G. McWilliam and R. A. Dewar, Nature (London), 182,
     1664 (1958).

54. V. Pretorius, Nature (London), 181, 177 (1958).

55. L. Onkiehong, Thesis, Technical Univ., Eindhoven, Holland,
     1960.

56. R. D. Condon, T. R. Scholly, and W. Averill, Intern. Symp.
     Gas Chromatogr., Edinburgh, 1960, preprints, p. N134.

57. J. C. Sternberg, W. S. Gallaway, and P. T. L. Jones, in
     Gas Chromatography, ISA Proceedings 1961 (N. Brenner,
     J. E. Callen, and M. D. Weiss, eds.), Academic Press,
     New York, 1962, p. 231.

58. L. S. Ettre, in Gas Chromatography (N. Brenner, J. E.
     Callen, and M. D. Weiss, eds.), Academic Press, New York,
     1962, p. 307.

59. L. S. Ettre, J. Chromatogr., 8, 525 (1962).

60. L. S. Ettre and H. N. Claudy, Chem. Can., 12, (9), 34 (1960).

61. A. J. Andreatch and R. Feinland, Anal. Chem., 32, 1021
     (1960).

62. P. Boček, J. Novák, and J. Janák, J. Chromatogr., 43, 431
     (1969).

63. P. Boček, J. Novák, and J. Janák, J. Chromatogr., 48, 412
     (1970).

64. P. Boček, J. Novák, and J. Janák, J. Chromatogr. Sci., 8,
     226 (1970).

65. C. H. Deal, J. W. Otvos, V. N. Smith, and P. S. Zucco,
     Anal. Chem., 28, 1958 (1956).

66. H. Boer, in Vapour Phase Chromatography (D. H. Desty, ed.),
     Butterworths, London, 1957, p. 169.

67. W. H. Graven, Anal. Chem., 31, 1197 (1959).

68. J. E. Lovelock and S. R. Lipski, J. Amer. Chem. Soc., 82,
     431 (1960).

69. J. E. Lovelock, Anal. Chem., 33, 162 (1961).

70. J. E. Lovelock, Nature (London), 189, 729 (1961).

71. P. F. Washbrooke, Chem. Z., 86, 377 (1962).

72. J. E. Lovelock, J. Chromatogr., 1, 35 (1958)

73. J. E. Lovelock, Nature (London), 181, 1460 (1958).

74. J. E. Lovelock, Intern. Symp. Gas Chromatogr., Edinburgh,
     1960, preprints, p. 9.

75. J. E. Lovelock, Nature (London), 182, 1663 (1958).

76. A. B. Littlewood, Gas Chromatography, Academic Press,
     New York, 1962, p. 272.

77. D. H. Desty, C. J. Geach, and A. Goldup, in Gas Chroma-
     tography (R. P. W. Scott, ed.), Butterworths, London,
     1960, p. 46.

78. J. E. Lovelock, in Gas Chromatography (R. P. W. Scott,
     ed.), Butterworths, London, 1960, p. 26.

79. I. E. Fowlis and R. P. W. Scott, J. Chromatogr., 11, 1
     (1963).

80. S. Ishii and B. Witkop, J. Amer. Chem. Soc., 85, 1832
     (1963).

81. H. T. Milles and H. M. Fales, Anal. Chem., 34, 860 (1962).

82. E. C. Horning, K. C. Maddock, K. V. Anthony, and W. J. A.
     VandenHeuvel, Anal. Chem., 35, 526 (1963).

83. L. D. Metcalfe, J. Gas Chromatogr., 1, 7 (1963).

84. J. Bohemen and J. H. Purnell, J. Appl. Chem. (London), 8,
     433 (1957).

85. J. S. Foster and J. W. Murfin, Analyst, 90, 118 (1964).

86. P. Boček and J. Novák, J. Chromatogr., 51, 375 (1970).

87. O. Hainová, P. Boček, J. Novák, and J. Janák, J. Gas
     Chromatogr., 5/8, 401 (1967).

88. R. S. Juvet, Jr. and J. Chiu, J. Amer. Chem. Soc., 83,
     1560 (1961).

89. J. Novák and J. Janák, J. Chromatogr., 28, 392 (1967).

90. A. Janik, J. Chromatogr., 54, 321 (1971).

91. A. Janik, J. Chromatogr., 54, 327 (1971).

92. A. Janik, J. Chromatogr., 64, 162 (1972).

93.   A. Janik, J. Chromatogr., 69, 321 (1972).

94.   A. P. Altshuller, J. Gas Chromatogr., 1, 6 (1963).

95.   E. R. Colson, Anal. Chem., 35, 1111 (1963).

96.   F. R. Cropper and S. Kaminski, Anal. Chem., 35, 735 (1963).

97.   K. Widmark and G. Widmark, Acta Chem. Scand., 16, 575
      (1962).

98.   R. Kaiser, Anal. Chem., 45, 965 (1973).

99.   W. J. Kirsten and P. E. Mattsson, Anal. Lett., 4, 235
      (1971).

100. A. Dravnieks, B. K. Krotoczynski, J. Whitfield, A. O'Don-
      nell, and T. Burgwald, Environm. Sci. Technol., 5, 1220
      (1971).

101. A. Dravnieks and A. O'Donnell, J. Agr. Food. Chem., 19,
      1049 (1971).

102. A. Zlatkis, H. A. Lichtenstein, and A. Tischbee,
      Chromatographia, 6, 67 (1973).

103. K. Grob and G. Grob, J. Chromatogr. Sci., 8, 635 (1971).

104. M. Novotný and A. Zlatkis, Chromatogr. Rev., 14, 1 (1971).

105. K. Grob and G. Grob, J. Chromatogr., 62, 1 (1971).

106. J. Novák, V. Vašák, and J. Janák, Anal. Chem., 37, 660
      (1965).

107. M. Selucký, J. Novák, and J. Janák, J. Chromatogr., 28,
      285 (1967).

108. J. Gelbičová-Růžičková, J. Novák, and J. Janák, J. Chro-
      matogr., 64, 15 (1972).

109. D. H. Desty, E. Glueckauf, A. T. James, A. I. M. Keulemans,
      A. J. P. Martin, and C. S. G. Phillips, in Vapour Phase
      Chromatography (D. H. Desty, ed.), Butterworths, London,
      1957, p. 11.

110. J. Gelbičová-Růžičková, J. Novák, and B. Chundela, Biochem.
      Med., 5, 537 (1971).

111. K. Grob, J. Chromatogr., 84, 255 (1973).

112. K. Grob and G. Grob, J. Chromatogr., 90, 303 (1974).

113. G. Machata, Mikrochim. Acta, 4, 691 (1962).

114. K. E. Matsumoto, D. H. Partridge, A. B. Robinson, L.
      Pauling, R. A. Flath, T. R. Mon, and R. Teranishi,
      J. Chromatogr., 85, 31 (1973).

115. M. Novotný, M. L. McConnell, M. L. Lee, and R. Farlow,
      Clin. Chem., 20, 1105 (1974).

116. V. V. Nalimov, The Application of Mathematical Statistics
      to Chemical Analysis, Pergamon, London, 1963, p. 39.

117. J. Janák, J. Chromatogr., 3, 308 (1960).

118. A. A. Zhukhovitskii and N. M. Turkel'taub, Gazovaya
      Khromatografiya, Gostoptekhizdat, Moscow, 1962, p. 241.

119. J. C. Giddings, Dynamics of Chromatography, Marcel Dekker,
      New York, 1965, p. 25.

120. J. M. Gill and H. W. Habgood, J. Gas Chromatogr., 5, 595
      (1967).

121. J. M. Gill and S. P. Perone, J. Chromatogr. Sci., 7, 709
      (1969).

122. A. I. M. Keulemans, Gas Chromatography, Reinhold, New
      York, 1957, p. 209.

123. S. Dal Nogare, C. E. Bennett, and J. C. Harden, in Gas
      Chromatography (V. J. Coates, H. J. Noebels, and I. S.
      Fagerson, eds.), Academic Press, New York, 1958, p. 117.

124. J. G. Karohl, J. Gas Chromatogr., 5, 627 (1967).

125. F. Bauman and F. Tao, J. Gas Chromatogr., 5, 621 (1967).

126. C. W. Childs, P. S. Hallman, and D. D. Perrin, Talanta,
      16, 629 (1969).

127. J. M. Gill, J. Chromatogr. Sci., 10, 1 (1972).

128. F. Caesar, Topics in Current Chemistry, Springer Verlag,
      Berlin, Heidelberg, New York, 1973, p. 139.

129. E. Proksch, H. Bruneder, and V. Granzner, J. Chromatogr.
      Sci., 7, 473 (1969).

130. J. Novák, K. Petrović, and S. Wičar, J. Chromatogr., 55,
      221 (1971).

131. H. Günsler, Chem. Ing. Tech., 42, 877 (1970).

132. A. W. Westerberg, Anal. Chem., 41, 1770 (1969).

133. R. Kaiser and M. Klier, Chromatographia, 2, 559 (1969).

134. F. Hock, Chromatographia, 2, 334 (1969).

135. H. A. Hancock, Jr., L. A. Dahm, and J. F. Muldoon,
     J. Chromatogr. Sci., 8, 57 (1970).

136. H. M. Gladney, B. F. Dowden, and J. D. Swalen, Anal.
     Chem., 41, 883 (1969).

137. S. M. Roberts, D. H. Wilkinson, and L. R. Walker,
     Anal. Chem., 42, 886 (1970).

138. A. H. Anderson, T. C. Gibb, and A. B. Littlewood,
     J. Chromatogr. Sci., 8, 640 (1970).

139. A. H. Anderson, T. C. Gibb, and A. B. Littlewood,
     Chromatographia, 2, 466 (1969).

140. A. H. Anderson, T. C. Gibb, and A. B. Littlewood, Anal.
     Chem., 42, 434 (1970).

141. W. J. Dixon and F. J. Massey, Introduction to Statisti-
     cal Analysis, McGraw-Hill, New York, 1957.

142. W. J. Youden, Statistical Methods for Chemists, Wiley,
     New York, 1951.

143. J. D. Hinchen, J. Gas Chromatogr., 5, 641 (1967).

144. W. E. Deming, Some Theory of Sampling, Wiley, New York,
     1950.

145. W. G. Cochran, Sampling Techniques, Wiley, New York,
     1953.

146. P. Boček, J. Novák, and J. Janák, J. Chromatogr., 42, 1
     (1969).

147. L. Mikkelsen, J. Gas Chromatogr., 5, 601 (1967).

148. J. M. Gill and C. H. Hartmann, J. Gas Chromatogr., 5,
     605 (1967)

149. G. Guiochon and M. Goedert, Chim. Anal. (Paris), 53,
     214 (1971).

150. V. Kusý, Anal. Chem., 37, 1748 (1965).

151. F. Bauman, F. Tao, and J. M. Gill, paper presented at
      the ACS meeting, New York, Sept. 1966.
152. Varian Aerograph, Previews & Reviews, August, 1967.
153. D. L. Ball, W. E. Harris, and H. W. Habgood, J. Gas
      Chromatogr., 5, 613 (1967).
154. T. A. Gough and E. A. Walker, J. Chromatogr., 45, 14
      (1969).
155. E. M. Emery, J. Gas Chromatogr., 5, 596 (1967).
156. F. Bauman, A. C. Brown, and M. B. Mitchell, J. Chromatogr.
      Sci., 8, 8 (1970).
157. N. Fozard, J. J. Frauses, and A. Wyatt, Chromatographia,
      5, 377 (1972).
158. R. A. Landowne, R. W. Morosani, R. A. Herrmann, R. H.
      King, Jr., and H. G. Sehnus, Anal. Chem., 44, 1961 (1972).

INDEX

A

Absolute-calibration technique, 74–77
calibration-curve method,
75, 76
direct-comparison method,
74, 75
Absolute-calibration technique, sample dilution,
98–100
calibration-curve method,
98, 99
direct-comparison method,
98
Accumulation, of gaseous components, 108–116
nonselective, 108–111
selective, 111–116; see
also Chromatographic-
equilibrium method

Accuracy, of results, 179
Analytical property, a, 13,
25, 27
additivity of, 29, 30, 44,
45
argon ionization detector,
63
auxiliary gas, a_α, 28, 29,
37
carrier gas, a_o, 13, 28–31,
43–45
carrier-gas/auxiliary-gas
mixture, $a_{o\alpha}$, 37
cross-section detector, 59
electron-capture detector,
61
flame ionization detector,
56
gas-density detector, 52

209

katharometer, 48
reference compound, a_r,
    43-45
reference-compound/car-
    rier-gas mixture, a_{ro};
    see Analytical property,
    change with composition
Scott's detector, 53
solute, a_i, 28-31, 43, 45
solute/carrier-gas mix-
    ture, a_{io}, 13, 28-31
solute/carrier-gas/auxil-
    iary-gas mixture,
    $a_{io\alpha}$, 37
stationary-phase vapor/
    carrier-gas mixture,
    a_{bo}; see Analytical
    property, change with
    composition
Analytical property, change
    with composition, 30ff
reference compound in ref-
    erence-compound/carrier-
    gas mixture, da_{ro}/dy_r,
    41-43
solute in solute/carrier-
    gas mixture, da_{io}/dy_i,
    30-35, 38-43
solute in solute/carrier-
    gas/auxiliary-gas
    mixture, $da_{io\alpha}/dy_i$, 37,
    38
stationary-phase vapor/
    carrier-gas mixture,
    da_{bo}/dy_b, 43
Automatic integration,
    165-169
disk integrator, 165, 166
electronic digital
    integrators, 166-169
Auxiliary gas, 36-40
effect on detector
    performance, 40
molar flow velocity of,
    dN_α/dt, 38
mole fraction of, in
    column-effluent/auxil-
    iary-gas mixture, y_α^*, 37
net response to solute,
    R_i^*, 38-40

volume flow velocity of,
    dV_α/dt, 28, 29, 38, 39
Auxiliary-gas reaction
    products, 38ff
molar flow velocity of,
    $dN_{\alpha p}/dt$, 39
volume flow velocity of,
    $dV_{\alpha p}/dt$, 38, 39

                    B

Background response, effect
    on net response; see
    Column bleeding, effect
    on response factors
Baseline correction; see
    Electronic integrators

                    C

Capacity ratio, $\underline{k}$, 7-9
Carrier gas, mole fraction in
    column-effluent/auxil-
    iary-gas mixture, y_o^*, 37
Chromatographic-equilibration
    method, 111-116
calculation of results,
    113-116
definition of, 111-112
degree of concentration,
    112, 113
partition coefficient in
    frontal and elution
    chromatography, 113,
    115
solute concentration in
    gas phase, q_{iG}, 114
solute concentration in
    sorbent, q_{iS}, 114
solute partition coeffi-
    cient in trapping tube,
    K_{iSG}, 113-115
solute retention volume
    in trapping tube, V_{Ri},
    115, 116
solute specific retention
    volume, V_{gi}, 116
solute weight in sorbent,
    w_{iS}, 114

solute weight in tube, W_{iT}, 113-116

solute weight in void space of tube, W_{iGT}, 114

void volume of trapping tube, V_{GT}, 113-116

Coefficient of variation; see Relative standard deviation, of results

Column bleeding, effect on response factors, 43

Column effluent, 19ff
  molar flow velocity of, dN/dt, 27, 32, 38
  mole number of, N, 22, 27
  volume of, V, 22, 23

Column length, $\underline{L}$, 9

Column radius, ρ, 9

Computer data-processing, 169-172
  comparison with integrators, 169
  hierarchical systems, 171, 172
  hybrid systems, 170
  pure digital systems, 171

Concentration-sensitive detectors, analytical aspects of, 18-24

Control substance, Z, 92

Controlled internal-normalization technique, 92-96; see also Internal normalization technique
  sample dilution, 103

Cross-sectional area of column, gas phase; see Gas phase cross-sectional area of column

D

Destructive detectors, analytical aspects of, 16-24

Detectors, 14ff
  characteristics of, 24

concentration-sensitive, 14; see also Concentration-sensitive detectors, analytical aspects of

concentration-sensitive/nondestructive, 21, 22

destructive, 14; see also Destructive detectors, analytical aspects of

differential, 14, 15

integral, 14, 15

mass-sensitive, 14; see also Mass-sensitive detectors, analytical aspects of

mass-sensitive/destructive, 23

nondestructive, 14; see also Nondestructive detectors, analytical aspects of

performance of, 28, 29, 31

Differential detectors, 14-16
  elution chromatography, 15, 16
  frontal chromatography, 15, 16

Dilution factor, 97

E

Effective carbon, C_{eff}, 57

Electronic integrators, 166-169
  baseline correction, 168, 169
  functioning of, 166-169
  logic of, 167, 168
  peak detection, 167, 168

Error propagation, 179
  relative standard deviation, 179
  standard deviation, 179

Errors, 177-198
  automatic peak-area measurement, 181-183; see also

Errors, electronic
integrators
instrumental contribution
of GC apparatus, 196
linear peak parameters,
196, 197
manual peak-area measure-
ment, 180, 181, 196,
197
peak-area determination,
comparison of methods,
183
peak height, 196, 197
relative molar response,
184, 197
volume handled in sample
preparation, 196, 197
volume of sample charge,
196, 197
Errors, conventional techniques, 184-198
absolute calibration,
184-188
comparison of, 198
internal normalization,
192-195
internal standardization,
188-190
standard addition, 190-192
Errors, electronic integrators, 182, 183
baseline-correction effect,
182
slope-detection effect, 182
Extraction, liquid samples,
117-138
direct analysis of extract,
118-121
double-extraction tech-
nique, 130, 131
double-sample standard-
addition technique,
124-128
elimination of system
factor, 121-130
multistep extraction, 136-
138
number of extractions,
136, 137

reference-mixture methods,
121-124
single-sample standard-
addition technique,
124, 125, 128, 129
solute concentration in
extract, q_{ie}, 117
solute concentration in
parent solution, q_{ip},
117
solute distribution between
parent solution and
extrahent, 117
solute partition coeffi-
cient, K_{ipe}, 117, 118,
122
solute recovery, 137, 138
solute weight in extract,
W_{ie}, 117
solute weight in parent
solution, W_{ip}, 117
standard-addition technique,
124-130
system factor of solute,
f_{ipe}, 118-121
system factor of standard,
f_{spe}, 119, 120
volume of extract, $V_{(i)e}$,
117
volume of extrahent, V_e,
117
volume of parent solution,
$V_{(i)p}$, 117
Extraction, liquid samples,
condensation of extract,
131-136
direct analysis of concentrate, 131, 132
reference-mixture methods,
133, 134
standard-addition tech-
niques, 134-136
Extraction, nonfluidic samples,
153, 154
comparison with liquid
samples, 154
solute partition coeffi-
cient, K_{ixe}, 154, 155
system factor, δ_{ipe}, 154,
155

F

Forward carrier-gas velocity, column outlet, u, 9
Forward carrier-gas velocity, within column, u_z, 8
Forward zone-velocity, within column, u_{iz}, 8

G

Gas-liquid systems, analysis of, 138-153
  analogy with liquid-extraction analysis, 140
  closed-loop stripping and trapping the components, 148-153
  comparison with liquid-liquid systems, 142
  complete solute stripping, 145-147
  density of liquid phase, d_L, 142
  effect of temperature, 140-142
  elimination of system factor, 140
  gas-phase recycling, 139, 148-153; see also Gas-liquid systems, analysis of, closed-loop head-space gas equilibration sampling
  solute activity coefficient, γ_{iL}, 141, 142
  solute concentration in gas phase, q_{iG}, 139
  solute concentration in liquid phase, q_{iL}, 139
  solute distribution between phases, 139
  solute equilibration, 143, 144
  solute partial pressure, P_i, 141
  solute partition coefficient, K_{iLG}, 139, 142

solute vapor pressure, P_i°, 141, 142
  solute weight in gas phase, W_{iG}, 139
  solute weight in liquid phase, W_{iL}, 139
  system factor, 139, 140
  volume of gas phase, V_G, 139
  volume of liquid phase, V_L, 139
  volume of sample withdrawn from gas phase, v_G, v_G', 144, 145
  weight of solute withdrawn from gas phase, w_{iG}, 143, 144
  weight of solute withdrawn from liquid phase, w_{iL}, 143
  weight of solute withdrawn from system, w_i, 143, 144
Gas-liquid systems, analysis of, closed-loop head-space gas equilibration sampling, 148-153
  gas-liquid solute partition coefficient, K_{iLG}, 149, 152
  gas-sorbent solute partition coefficient, K_{iSG}, 149, 150, 152
  gas volume, V_G^{++}, 149
  liquid volume, V_L, 149
  solute distribution in system, 149
  solute weight in liquid, W_{iL}, 149
  solute weight in sorbent, W_{iS}, 149, 150
  solute weight in trapping tube, W_{iT}, 150-152
  solute weight in void space of system, W_{iG}^{++}, 149
  solute weight in void space of trapping tube, W_{iGT}, 150

sorbent volume, V_S, 149
standard-addition
    technique, 152
system factor, f_{iLGS}, 149,
    150-152
Gas-liquid systems, gas-phase
    sampling from closed
    system, 139-145
double-sampling technique,
    145
single-sample standard-
    addition technique,
    143, 144
Gas-nonfluidic systems, analy-
    sis of, 153, 155, 156
closed-loop head-space gas
    equilibration sampling,
    156
double-sampling technique,
    156
solute distribution
    between phases, 155
solute partition coeffi-
    cient, K_{iXG}, 154, 155
solute stripping, 156
system factor, $\emptyset_{iXG}$, 154,
    155
Gas-phase cross-sectional area
    of column, ϕ, 8, 9
Gaussian curve, 10, 11, 158,
    174, 176
normalized, 158, 159
standard deviation of,
    length units, σ_b, 10, 11
standard deviation of,
    time units, σ_t, 11

H

Head-space gas analysis, non-
    fluidic samples, 155,
    156

I

Integral detectors, 14-16
elution chromatography,
    15, 16
frontal chromatography,
    15, 16

Integrators, 35, 36, 165-169
count number, η, 36
count number versus peak
    area, 36
count rate, $d\eta/dt$, 35, 36
disk and ball, 165, 166
electronic; see Electronic
    integrators
Interdependence of results,
    92, 193
Internal-normalization tech-
    nique, 90-96, 102, 103
controlled, 92-96
limitations of, 91, 92;
    see also Controlled
    internal-normalization
    technique
Internal-normalization tech-
    nique, sample dilution,
    102, 103
Internal standard, require-
    ments on, 81, 82
Internal-standardization
    technique, 77-82
calibration-curve method,
    79, 80
direct-comparison method,
    78, 79
Internal-standardization
    technique, sample
    dilution, 99, 100
calibration-curve method,
    100
direct-comparison method,
    99, 100

L

Linear relationship method,
    105

M

Mass-sensitive detectors,
    analytical aspects
    of, 18-24
Multiple extraction, see
    Extraction, liquid
    samples

N

Net response, see Response
Net response, peak maximum,
    R_{im}, 21
Nondestructive detectors,
    analytical aspects of,
    16-24

O

Overlapped peaks, 172-176
  curve fitting, 176
Overlapped peaks, linear
    separation of, 172-176
  democratic-distribution
    method, 173, 175
  perpendicular-drop method,
    173-175
  skimming method, 173, 175
  triangulation method, 173,
    175

P

Partition coefficient, solute
    in GC column, K, 2
Peak area, A, 34ff
  analytical significance of,
    34-36
  calculation of, from linear
    parameters, 158-161
  dependence on working
    parameters, 21-24
  flow dependence, 17, 20-24
  pressure dependence, 18-24
Peak area, calculation methods,
    159-161
  height × retention dis-
    tance, 160, 161
  height × width at mid-
    height, 159, 160
  triangulation, 160
Peak height, h_m, 10ff
  analytical significance of,
    161-164
  dependence on working
    parameters, 21-24

flow dependence, 17, 20-24
pressure dependence,
    18-24
relationship to peak area,
    158, 159
Planimetry, 157, 158
Plate height, H, 9, 163, 164
Plate number, $\underline{N}$, 9, 11, 21-23
Porosity of column packing,
    total, ε, 9
Pressure, in detector sensor,
    P, 18-20, 22, 24, 28,
    29, 32-39

R

Reaction products, 27ff
  molar flow velocity of,
    dN_p/dt, 27
  volume flow velocity of,
    dV_p/dt, 32
Recorder, chart speed, db/dt,
    11, 21-24, 36, 40,
    42-45
Recorder, instrument constant,
    β, 34-36, 39-45
Reference substance, rela-
    tive molar response,
    41
Reference substance, standard-
    addition technique,
    86-90
Relative molar response, RMR,
    40ff
  argon ionization detector,
    64, 65
  cross-section detector, 60
  definition of response
    factors, 42-45
  electron-capture detector,
    62
  flame-ionization detector,
    57
  gas-density detector, 52
  katharometer, 51
  peak area versus solute
    amount chromatographed,
    70, 72, 73

ratio of, standard to
  solute, RMR_{sr}/RMR_{ir}, 73
Scott's detector, 55
Relative standard deviation,
  of results, I_X, 178, 179
Reliability of results, factors influencing, 68-70
Repeatability of results, 178
Response, R, 13, 15, 25-28
  argon-ionization detector,
    63, 64
  carrier gas, R_o, 14
  cross-section detector, 59
  electron-capture detector,
    61, 62
  flame-ionization detector,
    56, 57
  gas-density detector, 52
  katharometer, 48-50
  Scott's detector, 53-55
  solute, R_i, 14, 21-24,
    28, 31-42
  solute/carrier-gas mix-
    ture, R_{io}, 14
Response factor, 23, 24, 42-45;
  see also Relative molar
  response
  concentration units, 42, 43
  determination without model
    compounds, 94, 95, 105
  homologous series, 44
Retardation factor, R, 8
Retention volume, solute in
  GC column, V_R, 9-12,
  21-23

            S

Sample, 67ff
  density of, 75
  effect of diluent, 103
  volume of, V, 72
  volume charged into gas
    chromatograph, v, 72
  weight of, W, 72
  weight charged into gas
    chromatograph, w, 72
Sensitivity attenuation, 71

Signal, S, 13, 17, 20, 25-28
  argon-ionization detector,
    63
  carrier gas, S_o, 14, 28
  cross-section detector, 59
  electron-capture detector,
    61
  flame-ionization detector,
    65
  gas-density detector, 52
  katharometer, 48
  Scott's detector, 53
  solute, S_i, 14, 18-20, 28
  solute/carrier-gas
    mixture, S_{io}, 14, 28
Solute, 2ff
  concentration profile of,
    in column effluent,
    10, 11
  distribution of, in GC
    phases, 7, 8
  instantaneous concentra-
    tion (wt./vol.) of, in
    column effluent, $c_i(t)$,
    10-12, 15, 16, 21-23
  mean concentration
    (wt./vol.) of, in GC
    zone at column outlet,
    c_i', 9-12
  mean concentration
    (wt./vol.) of, in GC
    zone within column,
    c_{iz}', 8
  molar flow velocity of,
    dN_i/dt, 32-34, 39-41
  molarity of, in column
    effluent, m_i, 33
  mole fraction of, in col-
    umn effluent, y_i,
    22-24, 28-32, 39
  mole fraction of, in col-
    umn effluent/auxiliary-
    gas mixture, y_i^*, 37, 38
  mole number chromato-
    graphed, N_i, 34-44, 70
  mole number of, in mobile
    phase, N_{im}, 7-9, 22, 23
  mole number of, in sorbent,
    N_{is}, 7, 8

partition coefficient of,
in GC system; see Par-
tition coefficient,
solute in GC column
peak-maximum concentration
(wt./vol.) of, at col-
umn outlet, c_i^*, 10-12,
21-23
peak-maximum concentration
(wt./vol.) of, versus
peak height, 21-23
relative concentration of,
in column effluent, 15,
18, 19; see also Solute,
mole fraction of, in
column effluent
sample volume charged into
gas chromatograph, $v_{(i)}$,
72
sample weight charged into
gas chromatograph, $w_{(i)}$,
72
total mole number of, in
GC zone, N_i, 8-12,
21-24; see also Solute,
mole number chromatographed
volume charged into gas
chromatograph, v_i, 72
volume of, in sample, V_i, 72
volume of the sample
handled in its prepar-
ation, $V_{(i)}$, 72
weight of, charged into gas
chromatograph, w_i, 72
weight of, in sample, W_i, 72
weight of the sample
handled in its prepar-
ation, $W_{(i)}$, 72
Solute, concentration in
sample, 71-73
molarity, m, 72
mole fraction, x, 72
weight fraction, g, 72
weight/volume, q, 72
Standard, 72 ff
mole number of, handled in
sample preparation, N_s,
72, 73

volume charged into gas
chromatograph, v_s, 72
volume of, in sample, V_s,
72
weight of, charged into
gas chromatograph, w_s,
72
weight of, in sample, W_s,
72
Standard, solution of, 72ff
volume of, charged into
gas chromatograph,
$v_{(s)}$, 72, 97
volume of, handled in
sample preparation,
$V_{(s)}$, 72, 97
weight of, charged into
gas chromatograph,
$w_{(s)}$, 72, 97
weight of, handled in
sample preparation,
$W_{(s)}$, 72, 97
Standard-addition technique,
82-90
calibration-curve method,
88-90
direct sample-charge
measurement, 83-85
reference-substance
methods, 86-88
Standard-addition technique,
sample dilution,
101, 102
calibration-curve method,
102
direct sample-charge
measurement, 101
reference-substance
methods, 101, 102
Standard deviation, of
results, S_x, 178, 179
Standard substance, 72, 73
Stationary phase, mole
fraction of, in column
effluent, y_b, 43; see
also Analytical prop-
erty, change with
composition

T

Temperature, in detector
     sensor, T, 22, 24, 28-39
Temperature-gradient tube,
     105, 109
Time interval of zone elution,
     at column outlet, Δt,
     9-11, 34-36
Time interval of zone elution,
     within column, Δt_z, 8
Trace analysis, 107ff
Trapping tube, 108ff
   capillary GC, 109, 110
   frontal chromatography,
       proceeding in, 111,
       112; see also Chromato-
       graphic-equilibration
       method
   GC determination of con-
       centrate, 110, 111
   head-space gas analysis,
       138, 139, 145, 147-153;
       see also Gas-liquid
       systems, analysis of
   liquid extraction of
       concentrate, 111
   sorbent volume in, V_s,
       114, 115
   sorbent weight in, W_s,
       116; see also Chromato-
       graphic-equilibration
       method
   sorption capacity of, 108
   transfer of concentrate to
       gas chromatograph, 108,
       109

void volume of, V_{GT}, **114,**
     116; see also Chromato-
     graphic-equilibration
     method

V

Volume, additivity of, 85
Volume, of detector sensor,
     V, 22, 24, 35-39
Volume, of mobile phase in
     GC column, V_m, 7
Volume, of sorbent in GC
     column, V_s, 7
Volume carrier-gas velocity,
     dV/dt, 9, 10
Volume column-effluent veloc-
     ity, 17, 20-24, 28-40
Volume of eluent, in GC zone
     at column outlet, ΔV,
     9
Volume of eluent, in GC zone
     within column, ΔV_z,
     8, 9

Z

Zone width, at column outlet,
     ΔL, 9
Zone width, within column,
     Δz, 8, 9